OCCUPATIONAL SAFETY AND HEALTH IN THE EMERGENCY SERVICES

Second Edition

James S. Angle

THOMSON

DELMAR LEARNING™   Australia Canada Mexico Singapore Spain United Kingdom United States

THOMSON

DELMAR LEARNING

Occupational Safety and Health in the Emergency Services, Second Edition

James S. Angle

Vice President, Technology and Trades SBU:

Alar Elken

Editorial Director:

Sandy Clark

Acquisitions Editor:

Alison Weintraub

Development Editor:

Jennifer A. Thompson

Marketing Director:

Dave Garza

Channel Manager:

Bill Lawrenson

Marketing Coordinator:

Mark Pierro

Production Director:

Mary Ellen Black

Production Editor:

Toni Hansen

Editorial Assistant:

Stacey Wiktorek

Library of Congress Cataloging-in-Publication Data:

Angle, James.
   Occupational safety and health in emergency services / James S. Angle.—2nd ed.
        p.   cm.
   Includes index.
   ISBN 1-4018-5903-8
   1. Emergency medical personnel—Health and hyygiene.   2. Fire fighters—Health and hyygiene.   3. Emergency medical personnel—Safety measures.   4. Fire fighters—Safety measures.   I. Title.

RA645.5.A54 2005
362.18—dc22

2004058079

NOTICE TO THE READER

*This book is dedicated to all those emergency responders
who have been injured or killed in the line of duty
and the valuable lessons they have taught us*

CONTENTS

FOREWORD

SECOND EDITION

For those of us who are engaged in the delivery of emergency services to our communities and companies, some new hazards and concerns have developed, in addition to our already difficult and dangerous work environment. To address these hazards, open this copy of Delmar's second edition of *Occupational Safety and Health in the Emergency Services* and look at the most up-to-date and easily understood principles of safety and health available.

From the early chapters concerning the history of emergency services health and safety, to the regulations and standards that we use to guide the health and safety programs for our emergency services personnel, this text covers it all. Chapters with titles that contain such phrases as *fire environment* safety, *medical and rescue* safety, and *specialized incident* safety are only a few of the unique areas that are explored and presented in this one-of-a-kind text. Whether you are a firefighter, emergency medical professional, or other specialized type of rescue worker, this new text has something for you. Not just interesting and applicable emergency scene tactics and procedures, but interesting and topical appendices on such subjects as firefighter fatalities, internet resources, 2-in/2-out regulations, and the initial report on the recent firefighter life safety summit.

Whatever your place in today's emergency services world, this text will be a guiding light toward safety and health for you and those who work with you, for you, and around you. Entry level emergency services personnel, emergency service supervisors and administrators of health and safety programs and departments will find this text a quality resource for just about every occupational safety and health issue that might arise in their areas of operations and responsibility. Be careful out there!

John Salka, Jr.
Battalion Chief
New York City Fire Department

FIRST EDITION

The book you are about to read is written about a significant change in our emergency response business that involves a major redefinition of how we manage the welfare and survival affairs of our human assets. We are now asking (actually requiring) our traditional systems, and the managers that go with those systems, to implement and refine an occupational health and safety program that effectively protects our workers.

This is a huge shift from the way we have done business in the past. In the old days we managed the troops a lot like we handled our apparatus and equipment. We purchased it, we put it in service, we used it, and when we were finished with it, we surplused it. In those days, the biggest difference between how we managed the humans and the "stuff" was that we actually spent a lot more time, effort, and resources maintaining the hardware than we did the humans. In fact, we were a lot more attracted to supporting the equipment because it did whatever we directed it to, it never talked back, and it always behaved in a way we could understand. For the most part, we used an old time vertical military model to manage our personnel. The model was basically designed to conduct the organizational activities of waging war. The model, and the mentality that went with it, assumed that if you were in the war business, some folks would get scuffed up (or worse). The system backend-loaded a benevolent set of responses that supported the basic fatalistic assumption of worker injury and death—things like big insurance plans, worker's compensation benefits, disability pensions, and huge funeral festivals. Those are (and probably always will be) very logical and legitimate benefits for workers who go into highly hazardous areas to do their jobs. The problem in the past was that these benefits were the major focus of how we responded to protecting the workers. These benefits in a well-managed system will actually compliment the occupational health and safety approach we use today.

The old backend-loading approach was directed toward reacting after something bad had happened. The new approach predicts and attempts to prevent the ugly stuff from ever occurring. Like any other organizational change, the new occupational injury/death prevention routine requires an old system to do new things, and it will take a while for us to become comfortable and skillful with the new program. This book provides a broad overview of the different parts of a complete health and safety program.

Implementing those program components provides a game plan for the reactive to proactive shift. This change is particularly difficult for us, simply because our historic approach is to react quickly to the customer's problem. Nothing in this book suggests that we change our commitment to the fast and effective delivery of service. That commitment is why we are and always will be in business. What this new routine does is create a sensible balance that also considers the responder as a customer.

This new balance recognizes that it is impossible for us to effectively deliver service if we are not protected from whatever is causing the customer a bad day. The health and safety program requires that we take a long-term, before/during/after, view of worker welfare.

In my travels, I hear many leaders say that their people are the most important asset in their organizations. Now when I hear someone say that, I think to myself, "OK, prove it." A workforce that routinely must put its bodies directly between the customers and their hazardous problems, has a very special set of welfare needs. How well we implement the program outlined in this book will determine in the most basic and important way if we can indeed "prove it."

Chief Alan V. Brunacini
Phoenix Fire Department

PREFACE

Each year, a number of emergency responders are killed or injured in the line of duty. It is my hope that this book will have a significant, positive impact on these statistics. Since the publication of the first edition much has changed for emergency service responders. More emphasis has been placed on preparedness for attacks using weapons of mass destruction. Technology has been improved in terms of tools and equipment used by responders. Additional standards and regulations have been promulgated in an effort to further the safety and health of responders. Other standards and regulations have been updated. While emphasis on occupational safety and health in fire and emergency medical services is a fairly new discipline, much can be learned by the concepts that have been in place in industry for many years. Increasing emphasis, and an increase in accountability for safe practices over the last 15 years addresses the need for education and training in safety and health practices that focuses on both fire and emergency medical services.

During the development of this second edition, a historic event took place. In Tampa, Florida, during March 2004, the National Fallen Firefighters Foundation hosted a 2-day symposium on preventing firefighter line-of-duty deaths. This symposium brought together 200 fire service leaders from across the country to discuss the issues of line-of-duty-deaths. The group looked at the line-of-duty-death problem from six perspectives: reducing fires, training, structure fires, vehicle safety, wildland, and wellness and fitness. After a day of working in subgroups, findings were presented to the full group. A summary will be published. This author had the honor, and considering the attendees, it was an honor, to attend and participate in this event. Further information from the symposium can be found in Appendix D.

ABOUT THIS BOOK

Since the safety trend has emerged, many colleges began to offer credit courses in this area as part of degree programs or as stand-alone courses. Additionally, this book is intended for such a course in emergency service occupational safety and health. The Fire and Emergency Services Higher Education group recommends an Emergency Safety and Health course as part of the associate degree curriculum. Further, fire and EMS departments can use this book as reference and for in-house instruction in safety and health programs. Students should also be encouraged to read emergency services trade journals as they frequently have articles focusing on safety and health issues.

This second edition continues with the comprehensive approach to emergency service occupational safety and health. The purpose of this book is to provide instructors, students, safety officers, fire and EMS department managers, and others in the emergency service field a one-stop resource for total safety program management.

The text introduces the reader to occupational safety and health from a historical standpoint, including a historical look at industrial safety and health. It then guides the reader through a process that will allow them to put the information to work within their organizations. The outcome, it is hoped, will be a safer work environment that will reduce costs associated with responder injuries and deaths and increase productivity.

HOW TO USE THIS BOOK

This text is designed to be used for a semester or quarter-hour college level course on occupational safety and health within the fire or EMS discipline. The book was also designed to be used by fire and EMS agencies, whether fully paid, combination, or volunteer, as an in-house reference, a manual for an organizational training program, or as a promotional testing reference. The text is designed to give the reader a broad overview of issues to consider for various types of incidents common to emergency response.

Chapter 1 provides a brief history of occupational safety and health for emergency services, with a brief examination into safety and health in industry over the years. It gives the reader an understanding of the importance and the rationale for the increasing emphasis over the past 15 years. Chapter 2 defines and reviews standards and regulations that have an impact on safety programs. Risk analysis and management is presented in Chapter 3 to allow the reader to form an understanding of this important concept before specific situations are presented. Specific safety issues including preincident, fire, medical/rescue, and specialized incidents, and postincident implications are described in Chapters 4 through 8. Chapter 9 focuses on issues relating to personnel roles and responsibilities. Personnel issues are often the most difficult to deal with in terms of safety programs. Chapters 10 through 12 deal with program management including development and ongoing management, evaluation of effectiveness, and information management. Chapter 13 includes a look at current, legal, ethical, and financial issues as well as future trends.

FEATURES OF THIS BOOK

The following features are included in this book to ensure that the contents of the chapters are reader-friendly and promote understanding about the importance of safety in the emergency services:

- Each chapter opens with a **Case Study** that is designed to create thought and stimulate discussion among readers. The stories are drawn from actual events and stress the importance of learning how to prevent accidents and fatalities and to advance optimum health and safety in the emergency services.

- **Learning Objectives** are clearly outlined at the beginning of each chapter to highlight what the student should expect to learn on completion of the chapter.
- **Key Terms** and **Notes** are integrated into each chapter to highlight important points and introduce new concepts.
- **Review Questions** based on the learning objectives presented at the beginning of the chapter are included at the end of each chapter, allowing readers to assess the knowledge they have gained.
- **Activities** at the end of each chapter are designed to allow readers to put into practice the knowledge they have gained through study of the concepts presented.
- **Additional Internet Resources** are located in the appendices and contain a listing of Web sites applicable to safety and health programs.

NEW TO THIS EDITION

This book was thoroughly updated with the latest information, practices, and trends to reflect the latest events within the industry, including the following:

- A revised account of the history of occupational safety and health provides a background and context for the reader and emphasizes the need to learn from past events in order to provide a safer environment for the future.
- It is up to date with the latest editions of the NFPA standards, including the *1500* series and the 2001 edition of *Standard 1710*, as well as the applicable OSHA standards and regulations.
- Case Studies are completely revised to reflect current events around the United States.
- The expanded content on terrorism reflects concerns of the post-9/11 era.
- A completely revised Chapter 13 focuses on current issues for safety and health programs, including legal, ethical, and financial and trends for the future.

SUPPLEMENT TO THIS BOOK

An *Instructor's CD-ROM* is available that contains many features to aid the instructor in classroom preparation:

- *Instructor's Guide,* including lesson plans and answers to review questions. Available in Microsoft Word so instructors may add their own notes or tailor lesson plans to meet the needs of their course.
- *PowerPoint Presentations* highlight the major concepts in each chapter and provide graphics and photos to visually enhance classroom presentation.
- *Chapter Quizzes* in Microsoft® Word format. Answers also provided.
- *FESHE and NFPA Correlation Grids* provide a quick reference for instructors.

Visit us at http://www.firescience.com for information on other fire science-related titles available through Thomson Delmar Learning.

ABOUT THE AUTHOR

James S. Angle is the fire chief of the Palm Harbor Fire Rescue Department in Pinellas County, Florida. Internationally accredited by the Commission on Fire Accreditation International, the department protects 60,000 people in a 20-square-mile area operating from four fire stations and provides a full range of services including fire prevention, public education, advanced life support first response, rescue, hazardous materials response, and fire suppression. He is a 29-year emergency service veteran, having begun his career in the Monroeville Fire Department, Company 4 in suburban Pittsburgh, Pennsylvania. He also served with Pittsburgh Emergency Medical Services for 5 years as a paramedic crew chief. His background includes employment with four fire rescue agencies both small and large, in the state of Florida including Hallandale Fire Rescue Department, Palm Beach County Fire Rescue, South Trail Fire Rescue, and Palm Harbor Fire Rescue. He has worked through the ranks from firefighter/paramedic to his current position as fire chief, including 8 years as a training and safety officer.

His education includes an associate degree in Fire Science Administration from Broward Community College, a bachelor's degree in Fire Science and Safety Engineering from the University of Cincinnati, and a master's degree in Business from Nova University. He also holds Instructor III certification from the Florida Bureau of Fire Standards and Training, is certified as a paramedic, and is a graduate of the Executive Fire Officer program of the National Fire Academy. He also holds Chief Fire Officer Designation from the Commission on Chief Fire Officer Designation.

As an instructor, he teaches various fire science classes for St. Petersburg College, including a three-credit occupational safety and health class. He has delivered seminars to national audiences at three Fire Department Instructors Conferences. He also publishes a monthly column in a national fire service newspaper.

ACKNOWLEDGMENTS

This section could be a chapter in and of itself. I would be remiss if I did not begin with thanks to my wife Joann, and my two sons, Anthony and Austin. When undertaking a project such as this, the support of the family is one of the most important needs. My lovely wife and great kids have always been there to lend a helping hand or provide some words of encouragement. For this, I once again give them my sincerest thanks. Without their support and willingness to give up Daddy for many weekends and evenings, this book would not have been possible. I would like to thank my father, who passed away during the writing of the first edition of this book, as he is the one who took me to my first fire, which at that moment made me realize that I never wanted to do anything other than work in this field.

I could name many individuals who had an impact on me throughout my career, but the list would be longer than the book. So many have helped me to make this my career and not just a job. I would like to recognize the Board of Fire Commissioners and the members of the Palm Harbor Fire Rescue Department, who have provided encouragement and support throughout this endeavor. I also thank the management and members of the South Trail Fire Department, where for 8 years I served as training and safety officer and where I realized that through a good process, management support, and member support, an organization can reduce injuries. As for the other departments in which I had the pleasure of serving, be assured that I learned something there, that I use it today, and it can be found somewhere in this book.

A special thanks to Mark Huth from Thomson Delmar Learning who took a chance on the first edition and allowed me to make a dream a reality. To Jeanne Mesick of Thomson Delmar Learning who during the writing of the first edition provided support, understanding, and friendship. For this edition, I would like to thank Alison Weintraub for accepting the proposal to write a second edition and for her support. To Jennifer Thompson as the developmental editor for all of her help, encouragement, and weekly phone calls to make sure that I was on track. Along with appreciation to all of the production staff at Thomson Delmar Learning, I would thank the reviewers whose time and comments ultimately made this a better book:

Gareth Burton
McDowell Owens Engineering
Kingwood, TX

Dave Casey, MPA, EFO, Bureau Chief
Bureau of Fire Standards, Training &
Safety
Florida Fire Division of State Fire
Marshal
Ocala, FL

Paul Grant, M.A. Ed
Red Rocks Community College
Lakewood, CO

Dennis Hopkins
Caldwell Community College
Hudson, NC

Gail Hughes, Assistant Professor
Chattanooga State Tech CC
Chattanooga, TN

Dr. Bill Lowe, E.M.T. – P
Clayton County Fire Department
Fayetteville, GA

William Shouldis
Director of Training
Philadelphia Fire Department
Philadelphia, PA

Deputy Chief James P. Smith
Philadelphia Fire Department
Philadelphia, PA

FIRE AND EMERGENCY SERVICES HIGHER EDUCATION (FESHE)

In June 2001, The U.S. Fire Administration hosted the third annual *Fire and Emergency Services Higher Education Conference*, at the National Fire Academy campus, in Emmitsburg, Maryland. Attendees from state and local fire service training agencies, as well as colleges and universities with fire related degree programs attended the conference and participated in work groups. Among the significant outcomes of the working groups was the development of standard titles, outcomes, and descriptions for six core associate-level courses for the *model fire science* curriculum that had been developed by the group the previous year. The six core courses are *Fundamentals of Fire Protection, Fire Protection Systems, Fire Behavior and Combustion, Fire Protection Hydraulics and Water Supply, Building Construction for Fire Protection,* and *Fire Prevention.*[1] The committee also developed similar outlines for other recommended courses offered in fire science programs. These courses included: *Fire Administration I, Occupational Health and Safety, Legal Aspects of the Emergency Services, Hazardous Materials Chemistry, Strategy and Tactics, Fire Investigation I,* and *Fire Investigation II.*

FESHE Content Area Correlation The following table provides a comparison of the twelve FESHE content areas with this text.

---

[1]*2001 Fire and Emergency Services Higher Education Conference Final Report*, (Emmitsburg, Maryland: U.S. Fire Administration) 2001, page 12.

FIRE AND EMERGENCY SERVICES HIGHER EDUCATION (FESHE) COURSE CORRELATION GRID

		Occupational Safety and Health in the Emergency Services, 2nd Edition Chapter Reference
Name:	*Occupational Safety and Health for the Fire Service*	
Course Description:	This course introduces the basic concepts of occupational health and safety as it relates to emergency service organizations. Topics include risk evaluation and control procedures for fire stations, training sites, emergency vehicles, and emergency situations involving fire, EMS, hazardous materials, and technical rescue. Upon completion of this course, students should be able to establish and manage a safety program in an emergency service organization.	
Prerequisite:	None.	
Outcomes:	1. Describe the history of health and safety programs.	1
	2. Identify occupational health and safety programs in industry today.	1
	3. Identify occupational health and safety programs for the emergency services.	1
	4. Describe the distinction between standards and regulations.	2
	5. Identify federal regulations that impact on health and safety programs.	2
	6. Identify the standards that impact on occupational health and safety.	2
	7. Identify the concepts of risk identification and risk evaluation.	3
	8. Describe the considerations for safety in fire stations and emergency response vehicles.	4
	9. Describe the components of an effective response safety plan.	4
	10. Describe the components of the preincident planning process.	4
	11. Describe the considerations for safety while training.	4
	12. Define the value of personal protective equipment.	5
	13. Describe the components of an accountability system in emergency operations.	5
	14. Define incident priorities and how they relate to health and safety.	5
	15. Describe the relationship of incident management as it relates to health and safety.	5
	16. Describe the methods of controlling hazards associated with responding to EMS, hazmat, and technical rescue incidents.	6, 7
	17. Explain the need for and the process used for postincident analysis.	8
	18. Describe the components and value of critical incident management programs.	8
	19. Describe the responsibilities of individual responders, supervisors, safety officers, and incident commanders, safety program managers, safety committees and fire department managers as they relate to health and safety programs.	9
	20. Describe the components of a wellness/fitness plan.	4

NFPA 1500 STANDARD CORRELATION GUIDE

NFPA 1500 Chapter		Occupational Safety and Health, 2nd Edition Chapter Reference
1	Administration	10
2	Referenced Publications	2
3	Definitions	
4	Fire Department Administration	3, 11, 12
5	Training and Education	4, 10
6	Fire Apparatus, Equipment, and Driver/Operators	4
7	Protective Clothing and Protective Equipment	5
8	Emergency Operations	5
9	Facility Safety	4
10	Medical and Physical Requirements	4
11	Member Assistance and Wellness Programs	4
12	Critical Incident Stress Program	8

*Also available on the accompanying Instructor's CD-ROM in electronic format along with correlation to NFPA standards 1403, 1404, 1561, 1581, 1582, and 1583.

Chapter

1

INTRODUCTION TO EMERGENCY SERVICES OCCUPATIONAL SAFETY AND HEALTH

Learning Objectives

Upon completion of this chapter, you should be able to:

- Discuss the history of occupational safety and health in industry.
- Discuss the history of emergency service safety and health programs.
- Identify, by using historical data, the safety and health problem as it is today.
- Describe the efforts that have been made to address the safety and health problem among emergency service occupations.
- List the national agencies that produce annual injury and fatality reports for emergency services.
- Identify the information that can be obtained from annual injury and fatality reports.

CASE REVIEW

A TALE OF TWO DEPARTMENTS

Department 1 currently has approximately 70 personnel and provides a full range of fire protection services including advanced life support at a first responder level. This department responds to about 5,600 alarms per year from four fire stations. In 1990, although the department had about half the staff that it has today, responded to fewer alarms, and provided emergency medical services (EMS) at a basic life support level, it recorded forty-four occupational injuries. In 1991, the injury rate rose to forty-six. In 1990 and 1991, the department began a comprehesive occupational safety and health program including assigning the duties of safety and health to the training officer and the formation of a safety and health committee. Together with support from all levels of the department, injury causes were identified, a safety audit was performed, and

programs put into place to make the workplace safer. By the year 2000, with alarms increasing each year, the injury rate was below ten.

Department 2 was a similar department of four stations, responding to 7,200 alarms each year. It too provided a full range of fire protection services, including advanced life support first response. In 1996, it experienced forty-one occupational injuries. Using some of the same processes as department 1 used, in 2002 this number was reduced to less than fifteen.

These two examples show that with a comprehensive health and safety program with support at all levels of the department, occupational injuries can be prevented. Using some the ideas and processes described in this book, one can be well on the way to making safety part of the department.

INTRODUCTION

Firefighting in the United States is sometimes described as the nation's most dangerous occupation because of the high rate of acute and chronic injuries, illness, and deaths. Firefighters are exposed to chemicals, carcinogens, extreme weather, building collapses, smoke, heat, physical and emotional stress, and traveling under less-than-ideal roadway conditions. It is no wonder that the occupation suffers a high rate of injury and death.

Emergency medical service responders face many of the same hazards as firefighters. In fact, emergency medical responders face additional exposure, including concerns such as infectious diseases. Most firefighters are also responsible for emergency medical response. In fact, the fire service is the largest provider of prehospital care.[1] Because of this close relationship between the firefighters and the responders to emergency medical incidents, a complete occupational safety and health program must focus on both fires and medical emergencies. Therefore, an examination of the problems and the methods to improve an historically poor record is conducted from the emergency service standpoint.

[1]Ludwig, Gary. *On Scene*, May 8, 1998. International Association of Fire Chiefs.

HISTORY OF OCCUPATIONAL SAFETY AND HEALTH IN INDUSTRY

occupational disease
an abnormal condition, other than an occupational injury, caused by an exposure to environmental factors associated with employment

occupational injury
an injury that results from exposure to a single incident in the work environment

Occupational safety and health in the workplace is certainly not a new concept, in fact the history can be traced back hundreds of years. Many important people throughout history have contributed to the study of **occupational diseases** and **occupational injuries**. Hippocrates (460–377 BC) described symptoms of lead poisoning among miners. Paracelsus (1493–1541) wrote a treatise on occupational diseases. He described lung diseases among miners and attributed the cause to vapors and emanation from metals. He also is known for his work in toxicology because of his observations of dose and response. Agricolos (1494–1555) wrote about the need to provide ventilation for miners. Bernardino Ramazzini (1633–1714) was known as the patron saint of industrial medicine. He had advised physicians to learn about occupational disease by studying the work environment and to always ask their patients their trade.

Over time, as these and many others were making strides in industrial occupational safety health, many events also occurred that highlighted the need for change in workplace safety. Often people had to fight for better and safer working conditions. These fights involved strikes, lockouts, fighting in the streets, and in some cases prison sentences. One famous event that most responders should be familiar with is the 1911 fire at the Triangle Shirtwaist Company in New York City. This fire claimed the lives of 146 workers and called attention to the use of sweatshoplike working conditions and lack of exits. In 1914, studies in New York and Ohio highlighted tuberculosis and unsanitary conditions leading to the abolishment of sweatshops. Studies in 1923 of "dusty trades" led to the development of industrial hygiene sampling equipment.

These notable events and studies led to the development of laws, regulations, and standards to better protect workers. Text Box 1–1 shows some significant milestones in the development of federal occupational safety and health legislation.

As a result of this history and the need for safe working conditions, a multitude of programs exist today in industry. These programs deal with air quality, hearing protection, and ergonomics, to name a very few. Many of these programs have been developed because they are required by regulations. Industrial facilities have safety and health divisions; colleges and universities grant degrees at every level for industrial hygiene and occupational safety and health. Regulations and standards set forth the minimum requirements for equipment and procedures. Facilities and equipment are inspected for safe operation, and personnel have access to medical surveillance in many occupations in order to find diseases early.

Emphasis on occupational safety and health in the emergency services is a new concept when considering the history of industrial safety and health described previously. However in the past 30 years it has been recognized that these age-old concepts too, must be applied to firefighters and EMS responders.

Text Box 1–1   *Important dates in the development of industrial safety and health.*

1908	Federal Workmen's Compensation Act—limited coverage
1916	Federal Highway Aid Act
1926	Federal Workmen's Compensation Act
1927	Federal Longshoremen's and Harbor Workers' Compensation Act
1936	Walsh-Healey Public Contracts Act
1952	Coal Mine Safety Act
1959	Radiation Standards Act
1960	Federal Hazardous Substances Labeling Act
1966	National Traffic and Motor Vehicle Safety Act Child Protection Act—banned hazardous household substances
1967	National Commission on Product Safety created
1968	Natural Gas Pipeline Safety Act
1969	Construction Safety Act Coal Mine Health and Safety Act
1970	Occupational Safety and Health Act
2003	Health Insurance Portability and Accountability Act (HIPAA)

HISTORY OF EMERGENCY SERVICES SAFETY AND HEALTH

For many years, the high injury and death rate for firefighters was accepted as part of the occupation (Figure 1–1). As firefighters moved into the role of EMS responders, the injury potential increased, not only because of additional exposures, but also because of increased incidents that led to a higher potential for responding and returning accidents.

However, fire and EMS occupational safety and health has matured over the 20 years, (see Text Box 1–2). Like in industry, the development of regulations and standards often came after a tragic event or through the study of occupational injuries and occupational diseases.

Many of the early fire service texts used in the education and training of firefighters and EMS responders make little or no reference to the injury and death problem from a standpoint of improvement. There are references to the dangers of the profession; however, little is provided for prevention strategies.

Figure 1–1 *Modern fire apparatus design has a focus on responder safety.*

Text-Box 1–2 *Important dates in the development of emergency service occupational safety and health.*

1960	International Association of Firefighter begins to publish a Firefighter Injury and Death Report
1973	*America Burning* published
1974	NFPA begins to publish an Annual Firefighter Injury report and an Annual Firefighter Death Report
1986	*NFPA 1403* adopted by NFPA
1987	*NFPA 1500* Adopted by NFPA
1996	OSHA rule on 2 In/2 Out
1997	Joint IAFC/IAFF Wellness/Fitness program developed
1998	NIOSH Fire Fighter Fatality Investigation and Program began
1998	OSHA Respiratory Protection Standard applied to firefighters
2001	*NFPA 1710* adopted

> The Commission on Fire Prevention and Control has made a good beginning, but it cannot do our work for us. Only people can prevent fires. We must become constantly alert to the threat of fires to ourselves, our children, and our homes. Fire is almost always the result of human carelessness. Each one of us must become aware—not for a single time, but for all the year—of what he or she can do to prevent fires.
>
> —President Richard M. Nixon
> September 7, 1972

Figure 1–2 *President Nixon's preface to America Burning.*

NFPA 1500

the National Fire Protection Association's consensus standard on fire department occupational safety and health

One of the first publicized documents making reference to firefighter safety was *America Burning*.[2] In this 1973 report to the president of the United States, the National Commission on Fire Prevention and Control reported on the nation's fire problem (see Figure 1–2). One section of the report focused on firefighter safety from the perspective of staffing, education, and equipment. The 1980s saw a significant increased interest in the safety problem facing the emergency responders, and this trend has continued into the 21st century. One of the most controversial, yet positive, steps in the safety and health area was the publication in 1987 of *NFPA 1500, Standard on Fire Department Occupational Safety and Health Program.* This standard and the series of other safety standards that followed set the safety and health movement in motion. The standard was controversial because of potential changes in operations and costs, but it caused managers, unions, volunteer fire chiefs, and others to note that the statistics for death and injury were unacceptable, and something had to be done.

NFPA 1403 Standard on Live Fire Training Evolutions was adopted in 1986, following a loss of life in fire training in 1982. This standard is the minimum standard for live fire training exercises, whether they be in acquired structures or at training centers. In 1996, the Occupational Safety and Health Administration made the ruling on the two-in/two-out procedures, which was followed 1998 by an updated respiratory protection standard. In 2001, *NFPA 1710* was approved by the National Fire Protection Association (NFPA). While primarily a deployment standard, many of the requirements dealt with issues relating to firefighter/EMS responder safety and health. These are described in more detail in the following chapters.

The past 10 years also saw the nation's responders becoming more prepared for the threat of terrorism and attacks by weapons of mass destruction. Following the September 11, 2001, attacks, fire and EMS providers had to step up their preparedness levels. These new threats to the safety and health of the responders caused shifts in thinking and highlighted the need to acquire new equipment and provide a new, higher level of training for this kind of event. The emergency ser-

---

[2]National Commission on Fire Prevention and Control, 1973. Washington, D.C.

vices were called on later in 2001 for anthrax events furthering the concern for the safety and health of responders from yet a different hazard.

Emergency services organizations are often reluctant to make and accept change. However, procedures, techniques, and often equipment do change. Often these changes are engineered without considering the human element of the change. For example, a department can implement the best all-hazard incident management system with scene accountability, but if individual companies or crews freelance with in the system, the program will fail. All too often, the attitude of the response personnel mimics that of our nation's population when it comes to fires and injuries—*it won't happen to me*. However, it *does* happen to someone almost 100,000 times a year for injuries and almost 100 times a year for deaths.

Since the publication of *NFPA 1500*, emergency services organizations have seen changes. It is now common to find organization charts that include a health and safety officer, new textbooks and new editions with sections on health and safety, safety and health committees formed with representation from all levels of the organization, and standard operational procedures intended to provide a safe working environment. Incident management systems have been established, the use of personal protective equipment is required, and safer equipment has been developed. All of this has been accomplished with the thought of improved safety and health for the responder. Many emergency service organizations offer yearly medical examinations in addition to the examination at time of hire. These regular physicals allow the formation of a baseline for an annual comparison of medical history over the length of a member's career. This trend will and should continue. Another significant improvement since the movement began is the recognition that the local injury problem should be evaluated. National statistics have been gathered about occupational injuries, but often this information is not compared to local statistics until a problem or tragic event occurs.

IDENTIFICATION OF THE SAFETY PROBLEM

A number of resources are available to help a safety and health officer develop a program. Clearly, local program design and development must focus on local problems. It would not make sense for Florida departments to invest considerable money to prevent injuries associated with exposure to cold. Likewise, a department providing EMS transport that experiences a high number of back injuries may want to make a substantial investment in a back injury prevention program. Local statistics can be gathered, and the data can be used to determine the local safety and health problem. Techniques for gathering this information and the process for the analysis are provided in Chapter 12. However, local data should be compared to that of the larger population from the state and national figures. The following organizations publish annual safety and health data that can be useful in this endeavor.

National Fire Protection Association

In addition to publishing standards, the NFPA publishes annual reports on both occupational injuries and deaths in the fire service in its *Fire Journal* magazine. The NFPA has been compiling these reports since 1974. The NFPA death survey is a report on all firefighter deaths and includes analysis of the fatalities in terms of type of duty, cause of death, age group comparisons, and population-served comparisons. The NFPA injury survey is not an actual survey of all departments but is a sample used to project the national firefighter injury experience. Although a prediction, the survey inspires a high level of confidence and is representative of all sizes and types of departments. These reports are very comprehensive and useful for the safety program manager to compare national to local experience.

United States Fire Administration

The **United States Fire Administration** (USFA) oversees the **National Fire Incident Reporting System** (NFIRS). Part of the NFIRS report is for collecting data for firefighter casualties. This data can be obtained through the USFA and can be used to compare national data to local data. However, there are some problems with this system. The NFIRS is a voluntary system in which departments may or may not participate, which creates the problem of not knowing what injuries or deaths may go unreported because a local jurisdiction did not participate in the program. Further, as the name implies, the report is used for incidents; therefore, injuries that occur to on-duty personnel outside of incidents is not captured.

The NFIRS data is formulated into reports in various formats and is available by mail from the USFA publication center. Much emergency service safety and health data, including case analyses of incidents in which fatalities or injuries occurred and protective clothing field test results, are also available from the USFA publication center (see Figure 1–3). The USFA also has an intensive worldwide Web page providing a means to order publications, but many are downloadable and can be reviewed immediately. The Internet address for the USFA can be found in Appendix II, or the USFA can be contacted by telephone at 1-800-238-3358.

International Association of Fire Fighters

Since 1960, the **International Association of Fire Fighters** (IAFF) has produced an annual firefighter injury and death study. This research reports on a number of issues relating to safety and health similar to that of the NFPA. However, the report is more in-depth and provides information on lost-time injuries and exposure to infectious diseases. This report is another resource for the safety and health program manager, but the data is only gathered from paid fire departments that have IAFF affiliation, which limits the study to these departments. The IAFF, along with the **International Association of Fire Chiefs** (IAFC), in 1997 presented the Fire

United States Fire Administration
department under the Federal Emergency Management Agency that directs and produces fire programs, research, and education

National Fire Incident Reporting System (NFIRS)
the uniform fire incident reporting system for the United States; the data from this report is analyzed by the United States Fire Administration

International Association of Fire Fighters (IAFF)
labor organization that represents the majority of organized firefighters in the United States and Canada

International Association of Fire Chiefs (IAFC)
organization of fire chiefs from the United States and Canada

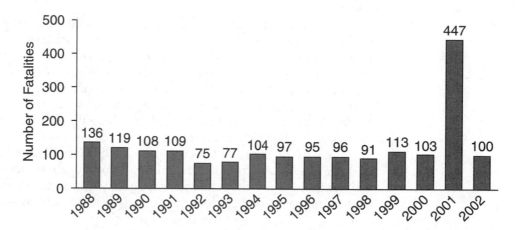

Figure 1-3
Firefighter fatalities 1988 to 2002 from the United States Fire Administration Report on Firefighter Fatalities 2002.

Service Joint Labor/Management Wellness-Fitness initiative. This initiative contains a component of data reporting to the IAFF on employee wellness, fitness, and injury issues. This report will be a great resource to the safety manager when sufficient data is received, but it will only contain data from departments that choose to participate.

Occupational Safety and Health Administration

Occupational Safety and Health Administration (OSHA)
federal agency tasked with the responsibility for the occupational safety of employees

The **Occupational Safety and Health Administration** (OSHA) is an agency within the Department of Labor. OSHA regulations are applicable to many public fire departments and to all private fire departments (see Chapter 2 for additional information on OSHA regulations applicable to public agencies in your state). OSHA requires certain reporting requirements for occupation-related injuries and deaths. This requirement for record keeping and reporting allows OSHA to compile useful statistics and to study causes of occupational injuries in order to develop prevention strategies and countermeasures. Because not every state requires that public fire departments comply with OSHA regulations, this data also is incomplete.

National Institute for Occupational Safety and Health

the National Institute for Occupational Safety and Health (NIOSH)
federal agency responsible for conducting research and making recommendations for the prevention of work-related injury and ilness; NIOSH is part of the Centers for Disease Control and Prevention (CDC)

Beginning in fiscal year 1998, the **National Institute for Occupational Safety and Health (NIOSH)** began the NIOSH Fire Fighter Fatality Investigation and Prevention Program to conduct independent investigations of firefighter line-of-duty deaths to formulate recommendations for preventing future deaths and injuries. The program does not seek to determine fault or place blame on fire departments or individual firefighters, but to learn from these tragic events and prevent future similar events. The goals of the program are to better define the magnitude and characteristics of line-of-duty deaths among firefighters, develop recommendations for the prevention of deaths and injuries, and to disseminate prevention strategies to the fire service. Another component of the program is the research database containing information for each injury incident. The database serves as a valuable tool for identifying trends and analyzing risk factors among line-of-duty injury deaths. Used in conjunction with individual incident reports, the database helps provide valuable information for developing broad-based recommendations for firefighter injury prevention programs. As with the investigation reports, personal and fire department identifiers are not included in the database.

One of the important goals for the NIOSH Fire Fighter Fatality Investigation and Prevention Program is disseminating investigative reports and other related publications to fire departments, firefighters, program planners, and researchers who can take action to help prevent firefighter line-of-duty deaths and injuries.

REVIEW OF CURRENT NATIONAL INJURY STATISTICS

To provide readers of this text with benchmarks and an insight into the type of information that can be obtained from an annual injury report, we examine the 2002 NFPA Firefighter Injury report. The NFPA report is usually published late in the year following the year studied. For example, the 2002 injury report would be published in late 2003. As described previously, this report is a statistical prediction of the national injury experience. The report examines the injury experience from a number of perspectives shown in the following lists. Each report is useful in analyzing and comparing the local problem with the national problem.

Injuries by Type of Duty The classification is divided into five subcategories (see Figure 1–4):

- Responding/returning
- Fireground
- Nonfire emergencies
- Training
- Other on-duty

In 2002, as in all other years, fireground injuries accounted for the highest number of injuries.

Nature of Injuries Nature of injury (see Figure 1–5) is subdivided into ten classifications, which are reported by type of duty as well.

- Burns (fire or chemical)
- Smoke or gas inhalation

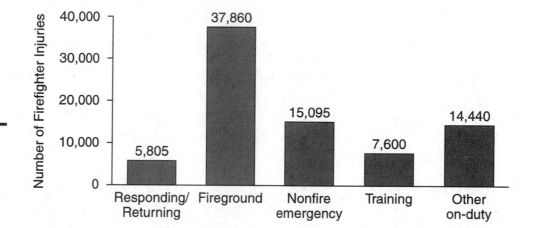

Figure 1–4
Firefighter injuries by type of duty, 2002. (With permission from NFPA.)

Nature of Injury	Responding to or Returning from an Incident		Fireground		Nonfire Emergency		Training		Other On-Duty		Total	
	Number	Percent	Number	Percent	Number	Percent	Number	Percent	Number	Percent	Number	Percent
Burns (fire or chemical)	145	2.5	3,205	8.5	175	1.2	300	4.0	30	0.2	3,855	4.8
Smoke or gas inhalation	95	1.6	2,245	5.9	110	0.7	60	0.8	65	0.5	2,575	3.2
Other respiratory distress	175	3.0	545	1.4	400	2.7	80	1.0	160	1.1	1,360	1.7
Burns and smoke inhalation	65	1.1	975	2.6	40	0.3	40	0.5	43	0.3	1,165	1.4
Wound, cut, bleeding bruise	1,375	23.7	8,215	21.7	2,545	16.9	1,325	17.4	2,760	19.1	16,220	20.1
Dislocation, fracture	190	3.3	985	2.6	315	2.1	305	4.0	545	3.8	2,340	2.9
Heart attack or stroke	120	2.1	345	0.9	165	1.1	50	0.7	340	2.4	1,020	1.3
Strain, sprain muscular pain	2,885	49.7	15,735	41.6	8,545	56.7	4,480	59.0	7,745	53.6	39,390	48.8
Thermal stress (frostbite, heat exhaustion)	175	3.0	2,415	6.4	105	0.7	375	4.9	155	1.1	3,225	4.0
Other	580	10.0	3,195	8.4	2,695	17.9	585	7.7	2,595	18.0	9,650	11.9
	5,805		37,860		15,095		7,600		14,440		80,800	

Note: If a firefighter sustained multiple injuries for the same incident, only the nature of the single most serious injury was tabulated.

Figure 1–5   *Nature of injuries by type of duty.* (With permission from NFPA.)

- Other respiratory distress
- Burn smoke inhalation
- Wound, cut, bleeding, bruise
- Dislocation, fracture
- Heart attack or stroke
- Strain, sprain, muscular pain
- Thermal stress (frostbite, heat exhaustion)
- Other

Fireground Injuries by Cause As fireground injuries historically cause the highest number of injuries per type of duty, fireground injuries by cause are also reported (see Figure 1–6). The eight categories for fireground injuries by cause are:

- Struck by object
- Stepped on, contact with object
- Extreme weather
- Exposure to fire products
- Exposure to chemicals or radiation
- Fell, slipped, jumped
- Overexertion
- Other

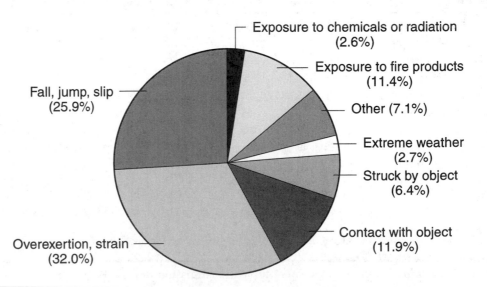

Figure 1–6
Fireground injuries by cause. (With permission from NFPA.)

Average number of fires, fireground injuries and injury rates by population of community protected, 2002

Population of Community Protected	Average Number of Fires	Average Number of Fireground Injuries	Number of Fireground Injuries Per 100 Fires	Number of Fireground Injuries Per 100 Firefighters
500,000 to 999,999	3,170.2	75.7	2.4	7.0
250,000 to 499,999	1,500.0	38.9	2.6	7.8
100,000 to 249,999	715.8	14.9	2.1	6.7
50,000 to 99,999	311.0	6.6	2.1	6.0
25,000 to 49,999	157.2	2.7	1.7	4.1
10,000 to 24,999	79.7	1.2	1.5	3.1
5,000 to 9,999	46.9	0.7	1.5	2.1
2,500 to 4,999	27.0	0.4	1.5	1.3
Under 2,500	13.5	0.2	1.5	1.1

Figure 1–7 *Average number of fires, fireground injuries, and injury rates by population of community protected.* (With permission from NFPA.)

■ **Note**
Examination of Figure 1–8 shows that while injuries have decreased in total, the rate of injuries per 1,000 fires has remained relatively constant.

■ **Note**
Review of this annual report must be required for safety and health program managers.

risk management
a process of using the available resources of an organization to plan and direct the activities in the organization so that detrimental effects can be minimized

Average Number of Fires and Fireground Injuries per Department by Population The average number of fires and fireground injuries per department by population for a community protected in 2001 is displayed in Figure 1–7. These results show that the number of fires a fire department responds to is directly related to the population protected and that the number of fireground injuries incurred by a department is directly related to its exposure to fires. The chart also shows the number of injuries per 100 firefighters.

Also of interest in the report is the 14-year report on total firefighter injuries per year. Examination of Figure 1–8 shows that while injuries have decreased in total, the rate of injuries per 1,000 fires has remained relatively constant. The NFPA report further summarizes the data presented in Figures 1–4 through 1–8 in the text of the report. The report also provides an overview and lengthy reports on selected incidents. Review of this annual report must be required for safety and health program managers.

WHAT IS BEING DONE

As described previously, a number of changes have occurred since the publication of NFPA 1500. Departments are taking a more proactive approach to safety and health. **Risk management** is now a concept as common in the fire station as it is

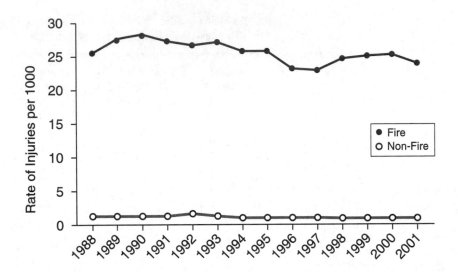

Figure 1–8 *Rate of injuries per 1,000 fire or nonfire incidents.*

in private industry. What has been done since NFPA 1500 can probably fit nicely into the acronym SAFEOPS, which means:

Supervision. Some departments have hired risk management specialists, assigned safety officers, empowered supervisors to be more safety aware, hired or contracted with fitness and wellness experts, and in many cases contracted with physicians to handle physicals and medical surveillance of department employees.

Attitude. Employees are made aware of the safety and health problem and empowered to be proactive concerning safety. When training begins, there is an emphasis placed on safety; in national journals, there are articles relating to safety and health almost monthly. By virtue of the resources put forth and the supervision focusing on safety and health, employees' attitudes toward safety are slowly improving. Attitude is the key to the success of any safety program; we must constantly seek improvement in attitude.

Fitness/wellness. Many departments have increased their commitment to fitness and wellness. Fire stations are equipped with workout equipment (see Figure 1–9) and in some cases departments are contracting with local gyms. Mental fitness has become a priority, with departments having critical incident stress debriefing teams and employee assistance programs. Some departments offer classes on weight reduction, smoking cessation, and eating right. Fitness and wellness are critical issues in the overall safety and health program.

Education. The level of safety education, both career and volunteer, for both entry-level and supervisory positions has increased, providing for a broad professional knowledge that improves the ability to collect, analyze, and present data related to problems. Programs have been developed to better prepare employees to safely perform their jobs, such as how to lift correctly to prevent sprains and strains.

Figure 1–9 *Fitness equipment should be provided in the station.*

NFPA 1403

National Fire Protection Association's consensus standard on live fire training evolutions

bloodborne pathogen disease carried in blood or blood products

Organizational involvement. Employee-based safety and health committees, fitness committees, and labor/management relations committees support the safety and health programs by developing new procedures or analyzing new equipment. Employee suggestion programs for safety-related improvements allow all members of an organization to provide input and be stakeholders in making their jobs safer.

Procedures. Operational procedures have been adopted or changed by many departments to reflect the needs of safety and health programs. Included are incident management systems, personnel accountability procedures, use of personal protective equipment, and responding to nonemergency incidents in a nonemergency mode.

Standards/regulations. Standards and regulations have also changed to meet the needs of the occupation. *NFPA 1403 Standard on Live Fire Training Evolutions* was written in hopes of preventing future deaths and injuries related to live fire training. Federal agencies also have promulgated regulations aimed at reducing occupational injuries and deaths. The OSHA **bloodborne pathogen** CFR 1910-1030 regulation requires procedures and precautions to prevent the spread of bloodborne diseases. Many departments may have adopted these procedures even if they had not been forced to comply.

This is just an overview of initiatives and programs that have been developed to improve the safety and health of emergency responders. Many are not universally accepted or practiced. Hopefully, this text will provide readers with further insight into some of these programs and procedures and motivate them to seek implementation in their own departments.

IS IT WORKING?

Even with the changes and emphasis on emergency service occupational safety and health, it is apparent by the statistics that the rate of illness, injuries, and fatalities has remained relatively constant. There are still a number of firefighter fatalities each year during training exercises. Both the fire service and EMS systems experience responder injuries and deaths during the responding and returning phase of the incident. Often these involve responders in their privately owned vehicles.

Does this mean that the efforts should be halted? Certainly not; in fact, it points out that the efforts must be continued and expanded. Every agency must make safety and health a top organizational priority. Solid research must be conducted to find out what is working and to develop new strategies to improve the safety and health of responders. This work has begun.

One example occurred in Phoenix, Arizona after the fire department experienced a line-of-duty-death. While the department had long prior developed procedures for the assignment of rapid intervention teams, is was not until after this tragic event, that the effectiveness of these teams was evaluated. Phoenix performed 200 rapid intervention drills as part of the recovery process following this fireground death. The results of the study were widely published and in many cases surprised fire service leaders. The results showed that rapid intervention is not rapid; in fact, in the drills, it took rescue crews an average of 2.50 minutes to get to a ready state. From the firefighter distress signal to the rapid intervention team entry was 3.03 minutes followed by 5.82 minutes for them to make contact with the downed firefighter. The average total time inside the building for each rapid intervention team was 12.33 minutes, with a average total time for rescue of 21 minutes. It was also found that it takes twelve firefighters to rescue one, one in five rapid intervention team members will get into some type of trouble themselves, and a 3,000-psi self-contained breathing apparatus (SCBA) bottle has 18.7 minutes of air (plus or minus 30%).

These kinds of hands-on studies are necessary to answer the question, "Is it working?" or more importantly "How well is it working?" New programs are being developed frequently to increase preparedness and to minimize to potential for injury and deaths among responders. Many of the concepts developed are presented in this text.

SUMMARY

Historically, prior to the publication of *America Burning*, little attention was given to overall safety and health in this profession. Occupational safety and health in industry can be traced back hundreds of years. However, for emergency service providers the emphasis is relatively new.

Emergency responders are in a hazardous profession. It is unlikely that given the nature of the occupation that all injuries and deaths could be totally eliminated. However, with good risk management the frequency and severities could be reduced. The first step in designing a safety and health program is to identify and understand the problem. This can be accomplished through analysis of local data and comparisons with the national experience. Several organizations provide national source for data relating to safety and health programs. The National Fire Protection Association publishes an annual report of firefighter injuries and firefighter deaths, the United States Fire Administration compiles injury and fatality data through the use of the National Fire Incident Reporting System, the International Association of Fire Fighters produces an injury and death report annually, and the Occupational Health and Safety Administration collects and analyzes data relating to occupational injuries. The National Institute for Occupational Safety and Health investigates firefighter line-of-duty deaths. All of these resources should be used to analyze the safety and health problems.

Many changes have occurred in the last 10 to 15 years regarding emergency responder safety, health, and wellness. The changes include increased awareness and priority on safety issues for front line supervisors, improvement in the safety attitude in all levels of the organization, increased emphasis on fitness and wellness, higher levels of education for entry-level employees and supervisors, involvement through committee and suggestion programs of all levels of the organization, revision and adoption of procedures related to safety operations, and promulgation of standards and regulations designed to improve the safety and health of responders. With all of these changes there is still significant progress that could be made. Through research, the development of new procedures and equipment will help to make the job of emergency service responder safer.

Concluding Thought: Improvements have been made to improve responder' safety and health. However, as represented in this chapter, the number of fires is down, EMS incidents remain on the increase, but the injuries have remained constant. Therefore, we need better programs, research, data, and organizational commitment to improve even more. Each organization can and should benchmark itself against similar organizations to further improve the statistics.

REVIEW QUESTIONS

1. What do the letters of the acronym SAFEOPS stand for?

2. Which of the following organizations survey every fire department, both paid and volunteer, in every state for injury and death statistics?

   a. USFA

   b. IAFF

   c. NFPA

   d. OSHA

   e. None of the above survey all departments

3. According to the text above, occupational safety and health moved ahead rapidly during which of the following decades?

    a. 1950s

    b. 1960s

    c. 1970s

    d. 1980s

4. Which major initiative did the IAFF and the International Association of Fire Chiefs work on together?

    a. *NFPA 1500*

    b. OSHA 1910.120

    c. The SAFEOPS approach

    d. A fitness/wellness program

Use Figures 1–4 through 1–8 to answer questions 5 through 10.

5. How many injuries per 1,000 fires occurred in 2001?

6. What percentage of injuries occurred on the fireground?

7. The most sprains and strains occurred during what type of duty?

8. What population range has the highest rate of injuries?

9. What population range has the highest rate of fires?

10. What caused the greatest percentage of fireground injuries?

ACTIVITIES

1. Obtain a copy of the most recent death or injury report from any of the organizations in this chapter. Review the statistics and compare them to those of your department.

2. Using the SAFEOPS approach, analyze what your department has done since the publication of the 1987 edition of *NFPA 1500* in terms of improving responder safety and health. List these approaches and evaluate their effectiveness.

Chapter 2

REVIEW OF SAFETY-RELATED REGULATIONS AND STANDARDS

Learning Objectives

Upon completion of this chapter, you should be able to:

- Discuss the difference between regulations and standards.
- Discuss the concept of standard of care.
- List and discuss federal regulations that have an impact on emergency responder safety and health programs.
- List and discuss the major National Fire Protection Association standards that have an impact on emergency responder occupational safety and health programs.
- Discuss the role of related regulations and standards and their safety and health implications.

CASE REVIEW

In July of 2002, a live-fire training exercise was being held in Osceola County, Florida. The live-fire exercise was being conducted in a vacant one-story, single-family dwelling constructed of cement block. During a simulated search and rescue exercise, two firefighters died as they searched for a mannequin dressed as a firefighter. After this incident occurred, the events leading up to this tragic event were investigated.

One of the agencies charged with investigating this incident was the Florida Division of State Fire Marshal. Its investigation concluded that a flashover occurred in the room of fire origin and that the deaths were the result of smoke inhalation and thermal injuries suffered during the training exercise. Findings also indicated that no written plan for the exercise was developed or reviewed by the department's administration prior to the drill. There was a lack of accountability being maintained at the point of entry to the fire room and there was too much fuel for the size of the room. The exit from the fire room was through two consecutive offset turns and then through a 26-inch opening, and through two other small rooms, which made it difficult for the maneuvering of personnel, and was a contributing factor to the accountability and accessibility for crews trying to attack the fire. Radio communications were reportedly less than optimal, and there did not appear to be a formal communications plan in place.

As part of the investigative report, recommendations were made to avoid such an event in the future. Among the recommendations was that the current edition of *NFPA 1403* should be adopted as state law by the legislature as part of the Firefighter Occupation Safety and Health Act. Furthermore, the Bureau of Fire Standards and Training was to develop a training program leading to a certification to be required for the lead instructor and safety officer for live-fire training.

Many of the contributing factors from this incident are identified and requirements set forth in *NFPA 1403*. It is important for the Safety and Health Officer in any organization to have a knowledge of applicable safety and health regulations.

Source: Poinciana Fire Investigation, Bureau of Fire Standards and Training, Florida Division of State Fire Marshal, 4/9/2003.

regulations
rules or laws promulgated at the federal, state, or local level with a requirement to comply

standards
often developed through the consensus process, standards are not mandatory unless adopted by a governmental authority

INTRODUCTION

The knowledge and understanding of safety-related standards and regulations is essential in all emergency service functions. "We have not adopted that standard" or "We are not an OSHA state" are not necessarily valid reasons to disregard the existence of standards and regulations.

There are differences between **regulations** and **standards**. Regulations are promulgated at some level of government by a governmental agency and have the force of law. Standards, sometimes called *consensus standards*, do not have the weight of law unless they are adopted as law by an authority having jurisdiction.

From the safety program management standpoint, an understanding of these applicable regulations and standards can be invaluable. These documents can

Code of Federal Regulations (CFR)
the document that contains all of the federally promulgated regulations for all federal agencies

■ Note

These regulations have the power of law and are enforced by the federal agency responsible for them.

consensus standards
standards developed by consensus of industry or subject area experts, which are then published and may or may not be adopted locally. Even if not adopted as law, these standards can often be used as evidence for standard of care

■ Note

Recommended practices are standards, which are similar in content to standards and codes, but are nonmandatory in compliance.

■ Note

Codes are standards that cover broad subject areas, which can be adopted into law independently of other codes, or standards.

actually provide a road map for an agency's safety and health program. This chapter focuses on the regulations and standards that apply to emergency service safety and health. Emphasis is placed on those that have the greatest effect and impact on the programs, but others are introduced also.

REGULATIONS VERSUS STANDARDS

Regulations

Regulations carry the weight of law and are mandatory in their requirements based on federal, and in some cases, state and local legislation. The entire collection of federal regulations is contained within the fifty titles of the **Code of Federal Regulations** (CFR). These mandatory requirements impact on emergency service safety and health programs, primarily those regulations found in Title 29 CFR, which is the OSHA regulations. These specific standards are discussed in further detail later in this chapter in the section Occupational Safety and Health Administration Regulations. Remember, these regulations have the power of law and are enforced by the federal agency responsible for them.

Some states have adopted federal laws, which then become mandatory state requirements. For example, the State of Florida Division of State Fire Marshal has adopted several OSHA regulations as they apply to firefighters and fire department employers including Section 1910.120, portions of 1910.134, 1910.146, and 1910.156 of Part 1910 of the Occupational Safety and Health Standards, 29 Code of Federal Regulations. In cases in which states adopt a regulation, state officials provide the enforcement.

Standards

Other published documents do not mandate compliance. These documents are commonly known as **consensus standards** because a group of professionals with a specific expertise have come together and agreed on how a specific task should be performed. The NFPA is one such standards-making group. The NFPA publishes standards, recommended practices, and codes. NFPA standards are documents in which the text includes the mandatory provisions of the requirements. Recommended practices are documents, which are similar in content to standards and codes, but are nonmandatory in compliance. Codes are standards that cover broad subject areas, which can be adopted into law independently of other codes, or standards.

The NFPA has no enforcement authority or power; its standards are considered advisory. However, a jurisdiction can adopt an NFPA standard and then the adopting authority would have legal rights to enforce the standard. The most common example of this process is the adoption of *NFPA 101®*, *Life Safety Code®*.

■ **Note**

Once a standard is adopted, the adopting governmental agencies have the power for enforcement.

standard of care
the concept of what a reasonable person with similar training and equipment would do in a similar situation

Both local and state governments have adopted the standard as part of the fire prevention program. Once a standard is adopted, the adopting governmental agencies have the power for enforcement.

The NFPA has developed a number of standards that impact safety and health programs. The most notable is *NFPA 1500, Standard on Fire Department Occupational Safety and Health Program.* Along with the *NFPA 1500* standard come a number of other standards in the *1500* series including medical requirements, infection control, and fire department safety officer. Further discussion of the NFPA standards is presented later in this chapter.

It is important to remember that these are consensus standards and are not mandatory unless adopted into law by local or state legislation. However, the fact that a group of professionals with related interest and expertise have come together and agreed on some minimum level of performance, these standards tend to become a **standard of care** in the particular subject.

STANDARD OF CARE

■ **Note**

The concept of standard of care is simple—everyone has certain expectations when it comes to performance.

personal alert safety system (PASS)
a device that produces a high-pitched audible alarm when the wearer becomes motionless for some period of time; useful to attract rescuers to a downed firefighter

The concept of standard of care is well known in the emergency medical field. From the first day of training, this concept is introduced to make students realize that to avoid liability they must perform in the same way as another reasonable person with the same training and equipment would perform. The concept of standard care is simple—everyone has certain expectations when it comes to performance. When a deposit is made to your checking account at the bank, you have a reasonable expectation that the teller will put the money into the right account. If mistakes are made, the bank has to take responsibility for the teller's actions.

This same concept can be applied to safety and health issues and closely relates to the existence of standards. The publication of a safety standard in and of itself has an impact on standard of care. For example, *NFPA 1500* requires that all responders to hazardous situations be equipped with a **personal alert safety system** (PASS). Should a firefighter get lost in a building fire and die, you may find yourself in court answering the question of why a PASS device was not provided. But NFPA is not law and has not been legally adopted in my jurisdiction you say. This answer will not be a viable defense when the firefighter's family sues for not following reasonable industry standards, because the standard of care has been defined by the NFPA document.

Standard of care is not a static concept but instead is very dynamic. It changes with new technologies and the development of regulations, standards, and procedures. Twenty-five years ago it was an acceptable standard of care to allow firefighters to fight fires without PASS devices. They did not exist and, if they did, no published document required them, therefore it was an acceptable practice.

OCCUPATIONAL SAFETY AND HEALTH ADMINISTRATION REGULATIONS

A number of OSHA regulations affect emergency service safety and health programs. Those most commonly used include confined space operations, respiratory protection, hazardous waste operations, and those regarding bloodborne pathogens. The following paragraphs describe each of these regulations in terms of applicability to safety and health programs and review the contents of each.

The application of OSHA regulations to public employers is a confusing issue. Under the Occupational Safety and Health Act of 1970, federal OSHA has no direct enforcement authority over state and local governments. However, a state may opt to implement its own enforcement program and may do so as long as federal OSHA approves the state's safety and health plan. A number of requirements, found in Section 18 of the Occupational Safety and Health Act, must be met for a state to have an approved plan. Some of the requirements include that the state be willing to assume responsibility for development and enforcement of occupational safety and health standards; assurances that the state will provide adequate funds for the administration and enforcement of the standards; that the state will, to the extent permitted by its law, establish and maintain an effective and comprehensive occupational safety and health program applicable to all employees of public agencies of the state and its political subdivisions; and that the state make the same reports to OSHA as it would if the state plan were not in effect.

There are currently twenty-six states or territories that have state OSHA plans (see Table 2–1). In these states, firefighters, including state, county, or municipal are covered by regulations promulgated by federal OSHA.

Table 2–1 *States and territories with OSHA plans.*

Alaska	Kentucky	New York*	Vermont
Arizona	Maryland	North Carolina	Virginia
California	Michigan	Oregon	Virgin Islands*
Connecticut*	Minnesota	Puerto Rico	Washington
Hawaii	Nevada	South Carolina	Wyoming
Indiana	New Jersey*	Tennessee	
Iowa	New Mexico	Utah	

*The plan covers public sector (State and local government) employment only.
Source: http://www.osha.gov.

OSHA 1910.146 (29 CFR 1910.146) Permit-Required Confined Spaces

This regulation applies to permitted confined space operations. There are several areas that apply to safety practices and procedures. The confined space regulation defines a confined space as any area that (1) is large enough and so configured that a person can bodily enter and perform assigned work; (2) has limited or restricted means for entry or exit (for example, tanks, vessels, silos, storage bins, hoppers, vaults, and pits are spaces that may have limited means of entry); and (3) is not designed for continuous employee occupancy (see Figure 2–1).

Clearly, there are cases when the emergency service employee would be required to enter the environments just defined. When private employers choose to use the local emergency service as their rescue team, 29 CFR 1910.146 (k) requires performance evaluation. Guidance for private employers who wish to use local rescue services can be found in 29 CFR 1910.146 Appendix F. Requirements contained in 1910.146 include those for written plans, atmosphere monitoring, notification, and equipment. Some companies that have permitted confined spaces on site have contracted with the local fire or rescue agency to perform services if needed. Even in the absence of this preevent agreement, we can expect that when an emergency occurs, the local emergency services will be summoned by a 911 phone call. From a safety program standpoint, these operations should not be performed until written procedures have been developed, the necessary equipment obtained, and the personnel have been properly trained.

■ Note

From a safety program standpoint, these operations should not be performed until written procedures have been developed, the necessary equipment obtained, and the personnel properly trained.

Figure 2–1 *Confined space entry is governed by OSHA regulations.*

OSHA 1910.134 (29 CFR 1910.134) Respiratory Protection

self-contained breathing apparatus (SCBA)
an atmosphere-supplying respirator for which the breathing air source is designed to be carried by the user

This regulation requires that the employer provide respirators when such equipment is necessary to protect the health of the employee (including **self-contained breathing apparatus** (SCBA)) (see Figure 2–2). The employer is required to provide the respirators that are applicable and suitable for the purpose intended. The employer is responsible for the establishment and maintenance of a respiratory protection program. The employee shall use the provided respiratory protection in accordance with instructions and training received. Requirements for a respiratory protection program are the following:

- Procedures for selecting respirators for use in the workplace
- Medical evaluations of employees required to use respirators
- Fit-testing procedures for tight-fitting respirators
- Procedures for proper use of respirators in routine and reasonably foreseeable emergency situations
- Procedures and schedules for cleaning, disinfecting, storing, inspecting, repairing, discarding, and otherwise maintaining respirators
- Procedures to ensure adequate air quality, quantity, and flow of breathing air for atmosphere-supplying respirators
- Training of employees in the respiratory hazards to which they are potentially exposed during routine and emergency situations
- Training of employees in the proper use of respirators, including putting on and removing them, any limitations on their use, and their maintenance
- Procedures for regularly evaluating the effectiveness of the program.

Figure 2–2 *SCBA usage and training falls under OSHA's respiratory protection regulation.*

Standard operating procedures (SOPs)
sometimes called standard operating guidelines, these are department-specific operational procedures, policies, and rules made to assist with standardized actions at various situations

Immediately dangerous to life and health
used by several OSHA regulations to describe a process or an event that could produce loss of life or serious injury if a responder is exposed or operates in the environment

Interior structural firefighting
the physical activity of fire suppression, rescue, or both, inside buildings or enclosed structures that are involved in a fire beyond the incipient stage

Most of these requirements should be incorporated into the organization's **standard operating procedures** (SOPs) also known as standard operating guidelines (SOGs). Immediately, even from these minimum requirements, one can see the safety program implications. One requirement in the respiratory protection regulation deals with fire scene staffing requirements. The procedures for operations in **immediately dangerous to life and health** (IDLH) environments are found in section 1910.134(g)(3) and further requirements for **interior structural firefighting** are described in 1910.134(g)(4). Summaries of the requirements of these two sections are listed in Text Box 2–1.

The requirements of 1910.134(g)(3) and 1910.134(g)(4) have come to be more commonly known as the two-in/two-out rule. Although the regulation does not preclude the firefighters from attempting a rescue in a known life-and-death situation (although when such a situation occurs, a written report should be completed and forwarded to the fire chief) prior to the rescue team being in place, it does require that before firefighters can operate in situations in which SCBA is worn four firefighters would be required. This ruling has implications to the operations of both career and volunteer fire service agencies as it applies to both. Appendix C contains an excellent and concise document issued by the IAFC and the IAFF that deals with this ruling.

The requirements of 1910.134 should be carefully evaluated by the safety and health officer prior to safety program development. Procedures that address operations when respirators are worn, staffing, the maintenance and repair of respirators, and the training that employees receive must be designed to meet this regulation.

OSHA 1910.120 (29 CFR 1910.120) Hazardous Waste Operations and Emergency Response (HAZWOPER)

The emergency service application of this regulation is clearly described as emergency response operations for releases of, or substantial threats of releases of, hazardous substances without regard to the location of the hazard. This OSHA regulation has a number of health and safety related requirements including a written safety and health program for employees involved in hazardous waste operations. This written plan must be made available to all employees and is required to contain, as a minimum, the following sections: (1) a means to identify, evaluate, and control safety and health hazards and provide for emergency response for hazardous waste operations; (2) an organizational structure; (3) a comprehensive workplan; (4) a site-specific safety and health plan, which need not repeat the employer's SOPs; (5) a safety and health training program; (6) a medical surveillance program; (7) the employer's SOPs for safety and health; and (8) any necessary interface between general program and site-specific activities.

The regulation also has specific requirements for personal protective equipment, training levels, and the necessary personnel to operate at an applicable incident.

Text Box 2–1   *Requirements of OSHA 1910.134(g)(3) and 1910.134 (g)(4).*

For all IDLH atmospheres, 1910.134(g)(3) requires that the employer shall ensure the following:

- One employee or, when needed, more than one employee is located outside the IDLH atmosphere.
- Visual, voice, or signal line communication is maintained between the employee(s) in the IDLH atmosphere and the employee(s) located outside the IDLH atmosphere.
- The employee(s) located outside the IDLH atmosphere are trained and equipped to provide effective emergency rescue.
- The employer or designee is notified before the employee(s) located outside the IDLH atmosphere enter the IDLH atmosphere to provide emergency rescue.
- The employer or designee authorized to do so by the employer, once notified, provides necessary assistance appropriate to the situation.
- Employee(s) located outside the IDLH atmospheres are equipped with:
  - Pressure demand or other positive-pressure SCBAs, or a pressure demand or other positive-pressure supplied-air respirator with auxiliary SCBA; and either
  - Appropriate retrieval equipment for removing the employee(s) who enter(s) these hazardous atmospheres where retrieval equipment would contribute to the rescue of the employee(s) and would not increase the overall risk resulting from entry; or
  - Equivalent means for rescue where retrieval equipment is not required under the previous

In addition to the requirements set forth under paragraph (g)(3), 1910.134(g)(4) requires that for interior structural fires, the employer shall ensure that:

- At least two employees enter the IDLH atmosphere and remain in visual or voice contact with one another at all times
- At least two employees are located outside the IDLH atmosphere
- All employees engaged in interior structural firefighting use SCBAs

  Note 1 to paragraph (g): One of the two individuals located outside the IDLH atmosphere may be assigned to an additional role, such as incident commander in charge of the emergency or safety officer, so long as this individual is able to perform assistance or rescue activities without jeopardizing the safety or health of any firefighter working at the incident.

  Note 2 to paragraph (g): Nothing in this section is meant to preclude firefighters from performing emergency rescue activities before an entire team has assembled.

Source: http://www.osha.gov

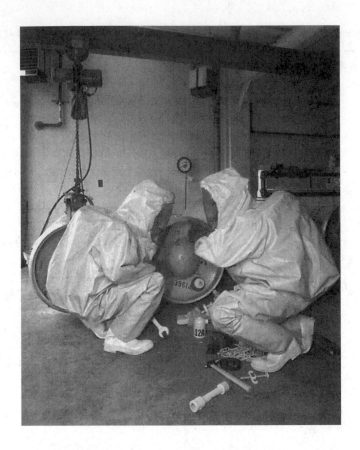

Figure 2–3 *The HAZWOPER regulations identify minimum requirements for hazardous material operations.*

For the safety program in a department that has the opportunity to respond to hazardous material incidents, familiarization and compliance with this regulation are necessary. The 1910.120 regulation was referenced in the January 1998 OSHA interpretation on staffing, specifically the requirement of personnel to utilize a buddy system (see Figure 2–3).

OSHA 1910.156 (29 CFR 1910.156) Fire Brigades

fire brigades
the use of trained personnel within a business or at an industry site for firefighting and emergency response

The OSHA fire brigade regulation applies to **fire brigades**, industrial fire departments, and private or contractual-type fire departments. The personal protective clothing requirements of this regulation apply only to those fire brigades that perform interior structural firefighting. The regulation specifically excludes airport crash fire rescue and forest firefighting operations. This regulation contains requirements for the organization, training, and personal protective equipment for fire brigades whenever they are established by employers.

The fire brigade regulation contains sections on (1) organizational statement (which establishes the existence of the brigade and personnel requirement),

(2) training and education, (3) firefighting equipment, (4) protective clothing, and (5) respiratory protection.

As with the 1910.134 and 1910.120 regulations, this regulation was also referenced in the January 1998 interpretation on staffing. Specifically section (f)(1)(i) was cited as it requires that SCBA be provided for all firefighters engaged in interior structural firefighting.

OSHA 1910.1030 (29 CFR 1910.1030) Occupational Exposure to Bloodborne Pathogens

The bloodborne pathogens regulation became effective on March 6, 1992. The intent of the regulation was to eliminate or minimize occupational exposures to hepatitis B virus (HBV), human immunodeficiency virus (HIV), and other bloodborne pathogens. The regulation is based on the premise that these exposures can be minimized or eliminated through a combination of engineering and work practice controls, personal protective clothing and equipment, training, medical surveillance, hepatitis B vaccination, signs and labels, and other provisions. The regulation applies to all employees who may have occupational exposure to blood or other potentially infectious material as defined in the regulation (see Figure 2–4).

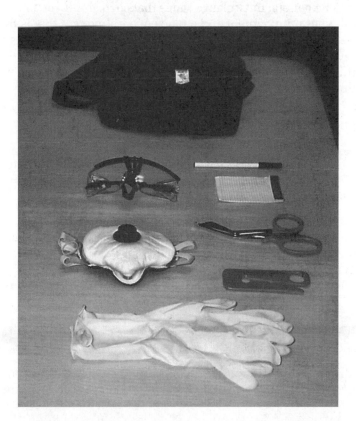

Figure 2–4 *Personal protective equipment is required to limit exposure to bloodborne pathogens.*

This regulation requires employers to have an exposure control plan that is reviewed and updated annually or sooner, as needed, to reflect new procedures or new equipment. Other requirements of the regulation include exposure determination, engineering and work practice controls, labeling of waste, housekeeping, personal protective clothing and equipment, record keeping, and decontamination procedures. The 1910.1030 regulation requires employers to provide hepatitis vaccinations to all employees with occupational exposure risk.

This regulation must be consulted when writing medical/rescue response procedures for the department's safety program, because almost all emergency service employees fall within the scope of this document in terms of occupational exposure.

OSHA General Duty Requirement

When the Occupational Safety and Health Act was enacted in 1970, it was clear to Congress that not all safety and health situations could be covered by the regulations. Knowing this, Congress included in the act the General Duty Clause, Section 5(a)(1).

The general duty clause states that each employer shall furnish to each of his employees employment and a place of employment that is free of recognized hazards that are causing or are likely to cause death or serious harm to his employees. Generally the OSHA Review Commission and court precedent have shown that the following elements would be necessary to prove a violation of the general duty clause:

- The employer failed to keep the workplace free of a hazard to which employees of the employer were exposed.
- The hazard was recognized.
- The hazard was causing or was likely to cause death or serious injury.
- There was a feasible and useful method to correct the hazard.

The implication of this clause to the safety and health program comes from a stated relationship with national consensus standards such as the NFPA. In the absence of a specific OSHA regulation, a national consensus standard (such as *NFPA 1500*) can be used as reference for guidance in enforcement of the general duty clause. This is an important consideration for the safety and health program manager when assessing the need to comply with consensus standards.

NFPA STANDARDS

With the publication of *NFPA 1500* in 1987 and the movement toward a greater concern for firefighter safety came the need to establish minimum performance criteria. To fulfill this need, the NFPA continued to develop other safety- and health-related standards in the *1500* series. These standards focused on infection control, fire department safety officers, medical requirements, and incident management systems. Because each of these has an impact on safety program management, each is described in this section.

NFPA 1500 Standard on Fire Department Occupational Safety and Health Program

Clearly the most significant document in the development of the firefighter safety movement was the 1987 adoption of *NFPA 1500*. This standard was adopted after having, at the time, the most comments in the history of the NFPA's standards-making process and was the first consensus standard for fire service occupation safety and health programs. It was thought that this standard would financially break fire departments and possibly put some out of business completely. In 1992 the second edition was adopted amid controversy surrounding safe fireground staffing. The *NFPA 1500* standard is on a five-year revision cycle with updated editions published in 1997 and 2002.

The *NFPA 1500* standard, 2002 edition, covers the essential elements of a comprehensive safety and health program and incorporates other standards into it as requirements. *NFPA 1500,* 2002 edition, lists fifty-two other documents that are incorporated by reference, forty-five of which are other NFPA standards. These referenced documents or portions of them are considered part of the *NFPA 1500* standard. See Figure 2–5.

NFPA 1500 provides the framework for any fire service-related safety and health program, including considerations for EMS response. The requirements outlined in *1500* are considered the minimum for an organization providing rescue, fire suppression, and other emergency service functions whether full or part time, as public, governmental, military, private, and industrial fire departments. The standard is not intended to apply to industrial fire brigades, which are covered in *NFPA 600, Standard on Industrial Fire Brigades*. Industrial fire brigades differ from industrial fire departments in that members of industrial fire departments have the primary responsibility for fire protection whereas industrial fire brigade members' primary responsibility is elsewhere in the facility, and they only handle fires as they first occur or are in the incipient stage. Again keep in mind that the standard is a minimum standard; nothing is intended to restrict the program manager to go beyond these requirements. The many facets of a safety program can be found in the twelve chapters of *NFPA 1500*. The appendices provide valuable guidance to the safety and health program manager. Text Box 2–2 lists the subject areas found in *NFPA 1500*.

NFPA 1500 provides the user with comprehensive direction for developing a safety and health program. Even if not adopted, this standard is known throughout the emergency field and is accepted as a standard of care. No safety program manager should be without a copy of *1500* and its referenced documents.

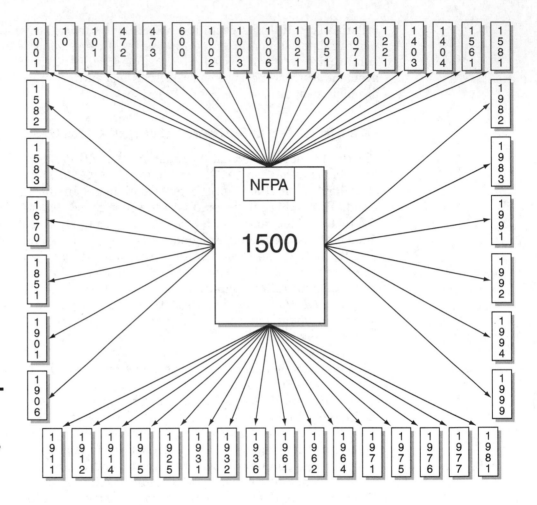

Figure 2–5 *The NFPA 1500 standard directly references forty-five other NFPA standards.*

Text Box 2–2 *Safety and health subjects in NFPA 1500.*

Fire Department Administration

Training and Education

Fire Apparatus, Equipment, and Driver/Operators

Protective Clothing and Protective Equipment

Emergency Operations

Facility Safety

Medical and Physical Requirements

Member Assistance and Wellness Programs

Critical Incident Stress Program

NFPA 1521 Standard for Fire Department Safety Officer

With the 1992 edition of *Standard for Fire Department Safety Officer*, the NFPA changed the standard number from *1501* to *1521* to better fit the numbering plan for fire service occupational safety and health documents. In 1997, the standard was updated to better reflect the difference between a health and safety officer and an incident safety officer. This standard contains minimum requirements for the assignment, duties, and responsibilities of a health and safety officer and an incident safety officer for a fire department or other fire service organization (Figure 2–6).

The functions of a fire department's health and safety officer (HSO) are defined in Chapter 5 and incident safety officer (ISO) in Chapter 6 of the standard. The fire chief appoints the HSO, while the ISO is a qualified member who works directly for the Incident Commander (IC) as part of the command staff. Table 2–2 compares the various responsibilities and functional areas as required by *NFPA 1521*. The organization's safety program manager should meet the requirements of this standard and ensure that assigned incident safety officers also meet the standard.

NFPA 1561 Standard on Emergency Services Incident Management System

NFPA 1500 requires that fire scene operations be conducted using an Incident Management System (IMS). *NFPA 1561* contains the minimum requirements for an IMS that can be used for all emergency incidents with the purpose of defining and describing the essential elements. To this end, Chapter 4 of the standard describes the system structure, including implementation, communications,

Figure 2–6 *A safety officer should be in place throughout the incident.*

Table 2–2 *Various responsibilities and functional areas as required by NFPA 1521.*

Health and Safety Officer	Incident Safety Officer
Risk management	Incident management system
Laws, codes, and standards	Incident scene safety
Training and education	ISO for fire suppression incidents
Accident prevention	ISO for emergency medical service operations
Accident investigation, procedures, and review	ISO for hazardous materials operations
Records management and data analysis	ISO for special operations
Apparatus and equipment for compliance with standards	Accident investigation and review
Facility inspection	Postincident analysis
Health maintenance	
Liaison for safety and health committee	
Occupational safety and health committee	
Infection control	
Critical incident stress management	
Postincident analysis	

multiagency incident response, command structure, training and qualifications, resource accountability, personnel accountability, incident scene rehabilitation, incident scene rehabilitation tactical level component, and evaluation and triage of emergency responder injuries.

Chapter 5 of the standard describes the system components, the incident commander, command staff, information officer, liaison officer, incident safety officer, general staff, operations functions, staging, planning functions, logistics functions, finance/administration, and supervisory personnel

The safety program manager will find this document very helpful when developing procedures for and training on incident command.

NFPA 1581 Standard on Fire Department Infection Control Program

NFPA 1581 was developed to reduce the exposures to infectious diseases by responders in both emergency and nonemergency situations and is compatible with the guidelines developed by the Centers for Disease Control and Prevention described later in this chapter. The standard defines the minimum requirements for fire department infection control programs. It begins with a description and requirements for the program components including risk management, training and education, infection control officer, immunization and testing, and exposure incidents. Chapter 3 sets forth the requirements for the department's facilities such as disinfecting areas, cleaning areas, and storage rooms. Chapter 4 describes requirements related to fire department apparatus used to transport patients. Chapter 5 covers emergency medical operations including personnel issues, infection control garments and equipment, and the handling of sharp objects. Cleaning, disinfecting, and disposal issues are addressed in Chapter 6; these are skin washing, disinfectants, emergency medical equipment, clothing, and disposal of material. This standard is of value as a resource to assist in compliance with the OSHA 1910.1030 regulation.

NFPA 1582 Standard on Comprehensive Occupational Medical Program for Fire Departments

NFPA 1582 was developed to cover the medical requirements necessary for persons engaged in rescue, fire suppression, emergency medical services, hazardous materials mitigation, special operations, and other emergency service functions. The requirements in this standard are intended to apply to both existing employees and to candidates. There are two categories of medical conditions in the standard, A and B. Persons with category A conditions are not allowed to perform emergency operations or training functions. Category B conditions are to be evaluated on a case-by-case basis with the fire department physician. This standard contains the medical requirements for firefighters whether full or part time, paid or volunteer, and specifies the minimum medical requirements.

The subjects covered in *NFPA 1582* include roles and responsibilities, essential job tasks, medical evaluations of candidates, occupational medical evaluation of current members, annual occupational fitness evaluation of current members, and essential job tasks in specific evaluation of medical conditions. This standard is helpful in determining the frequency of physical exams, developing criteria for the physical exam, and providing direction to a department physician as to what conditions may disqualify a person from performing emergency service duties.

NFPA 1583 Standard on Health-Related Fitness Programs for Fire Fighters

The focus during the development of this standard was the intent that a firefighter or other emergency service responder should have a comprehensive health program that includes a fitness component. The standard can be used as a companion document to *NFPA 1582*. This standard can also be used in conjunction with the Joint Labor-Management Wellness Initiative developed by the IAFF and the IAFC, which is discussed later in this text.

The requirements in this standard are the minimum requirements of the development, implementation, and management for a health-related fitness program. Subjects include the duties of the health and fitness coordinator, fitness assessments, exercise and fitness training programs, health promotion education, and data collection. The organization's safety and health officer will find this information useful in adding the fitness component to the complete safety and health program.

NFPA 1584 Recommended Practice on the Rehabilitation of Members Operating at Incident Scene Operations and Training Exercises

While not a standard, the NFPA has published a recommended practice for incident scene rehabilitation and training. This recommended practice describes requirements for medical evaluation and treatment, food and fluid replenishment, relief from climatic conditions, rest and recovery, and accountability. *NFPA 1584* covers preincident considerations such as having SOPs in place for rehabilitation and training personnel in heat and cold stress. Chapter 5 of the standard covers the requirements for a rehabilitation area including site characteristics; Chapter 6 describes the steps necessary for the process at the incident scene of training site; and Chapter 7 deals with postincident recovery and watching for signs of dehydration. This recommended practice is helpful in developing rehabilitation procedures. The need for fireground rehabilitation is described further in Chapter 5 of this text.

NFPA 1710 Standard for the Organization and Deployment of Fire Suppression Operations, Emergency Medical Operations, and Special Operations to the Public by Career Fire Departments

NFPA 1710 was first published in 2001 after a great deal of controversy and public comment. This standard contains requirements relating to the organization and deployment of fire suppression operations, emergency medical operations, and special operations to the public by career fire departments. One of the purposes stated in this standard is to specify the minimum criteria addressing the career public fire suppression operations, emergency medical service, and special operations with regard to the occupational safety and health of fire department members.

Chapter 4 of the standard contains the requirements for the fire department organization. The extensive Chapter 5 covers the requirements for fire department services, including fire suppression, emergency medical operations, special operations, airport rescue and firefighting, marine rescue and firefighting, and wildfire incidents. Each of these sections has detailed minimum requirements for staffing, deployment, and operations. Chapter 6 describes the requirements for fire department systems including safety and health, communications, training, preincident planning, and incident management.

This standard not only impacts fire department organization and deployment but also has a number of safety and health program implications.

NFPA 1720 Standard for the Organization and Deployment of Fire Suppression Operations, Emergency Medical Operations, and Special Operations to the Public by Volunteer Fire Departments

Similar in content and structure to *NFPA 1710*, this standard contains minimum requirements relating to the organization and deployment of fire suppression operations, emergency medical operations, and special operations to the public volunteer fire departments. In contrast to specific requirements in *1710*, *NFPA 1720* considers the uniqueness of the volunteer fire service, the different services they provide, and how they deploy and respond. For this reason, many of the specific requirements are left up to the local authority having jurisdiction to determine.

Chapter 4 of the standard is fire department organization, operations and deployment, including sections about fire suppression organization and operations, intercommunity organization, emergency medical services, and special operations. Chapter 5, like *NFPA 1710*, describes the requirements for fire department systems.

Safety and health officers in volunteer organizations should be knowledgeable in this standard and work with administrators to set the minimum standards for the deployment and operations as they relate to the safety and health program.

Other NFPA Standards

There are forty-five other NFPA standards incorporated by reference into *NFPA 1500*, 2002 edition. Four of these standards, *NFPA 1561, NFPA 1581, NFPA 1582*, and *NFPA 1583*, are from the *1500* series and were described in the preceding section. The other 41 are described in this section. They are standards that affect safety and health from other standpoints, for example, apparatus specifications, live fire training, and performance requirements for protective clothing. It is important to recognize the existence and inclusion of these standards into *NFPA 1500*.

> *NFPA 10 Standard for Portable Fire Extinguishers* applies to the selection, installation, inspection, maintenance, and testing of portable extinguishing equipment.
>
> *NFPA 101® Life Safety Code®* standard identifies the minimum life safety requirements for various occupancies.
>
> *NFPA 472 Standard for Professional Competence of Responders to Hazardous Materials Incidents* identifies the competencies for first responders at the awareness level, first responders at the operational level, hazardous materials technicians, incident commanders, and off-site specialist employees.
>
> *NFPA 473 Standard for Competencies for EMS Personnel Responding to Hazardous Materials Incidents* identifies the requirements for basic and advanced life support personnel in the prehospital setting.

NFPA 600 Standard on Industrial Fire Brigades identifies minimum requirements for organizing, operating, training, and equipping industrial fire brigades.

NFPA 1001 Standard for Fire Fighter Professional Qualifications identifies performance requirements, specifically the minimum requirements for firefighters.

NFPA 1002 Standard for Fire Apparatus Driver/Operator Professional Qualifications identifies minimum job performance for the firefighter driver/operator of fire department vehicles.

NFPA 1003 Standard for Airport Fire Fighter Professional Qualifications identifies the minimum job performance for the airport firefighter.

NFPA 1006 Standard for Rescue Technician Professional Qualifications establishes the minimum job performance requirements necessary for fire service and other emergency response personnel who perform technical rescue operations.

NFPA 1021 Standard for Fire Officer Professional Qualifications identifies the performance requirements necessary to perform the duties of a fire officer and establishes four levels of progression.

NFPA 1051 Standard for Wildland Fire Fighter Professional Qualifications identifies the minimum job performance requirements for wildland fire duties and responsibilities, including wildland firefighter I and II as well as wildland fire officer I and II.

NFPA 1071 Standard for Emergency Vehicle Technician Professional Qualifications identifies the minimum job performance requirements for a person to be considered qualified as an emergency vehicle technician. The standard applies to personnel who are engaged in the inspection, diagnosis, maintenance, repair, and testing of emergency response vehicles.

NFPA 1221 Standard for the Installation, Maintenance, and Use of Emergency Services Communications Systems covers the installation, performance, operation, and maintenance of public emergency service communications systems and facilities.

NFPA 1403 Standard on Live Fire Training Evolutions requires the establishment of procedures for training of fire suppression personnel engaged in structural firefighting under live fire conditions (see Figure 2–7).

NFPA 1404 Standard for Fire Service Respiratory Protection Training contains the minimum requirements for the training component of the Respiratory Protection Program as required by *NFPA 1500*.

Figure 2–7 *NFPA 1403 defines minimum requirements for live fire training exercises.*

NFPA 1670 Standard on Operations and Training for Technical Search and Rescue Incidents identifies and establishes levels of functional capability for safely and effectively conducting operations at technical rescue incidents.

NFPA 1851 Standard on Selection, Care, and Maintenance of Structural Fire Fighting Protective Ensembles specifies the minimum selection, care, and maintenance requirements for structural firefighting protective ensembles and the individual ensemble elements that include coats, trousers, coveralls, helmets, gloves, footwear, and interface components to be compliant with *NFPA 1971.*

NFPA 1901 Standard for Automotive Fire Apparatus applies to all new automotive fire apparatus. The 1996 edition combined *NFPA 1902,*

1903, and *1904* into one standard to cover all types of fire department apparatus as opposed to having a different standard for each.

NFPA 1906 Standard for Wildland Fire Apparatus applies to new automotive fire apparatus designed primarily to support wildland fire suppression operations.

NFPA 1911 Standard for Service Tests of Fire Pump Systems Fire Apparatus applies to the service testing of fire pumps and attack pumps of fire department automotive apparatus excluding apparatus equipped solely with pumps rated at less than 250 gpm (950 L/min).

NFPA 1912 Standard for Fire Apparatus Refurbishing specifies the minimum requirements for the refurbishing of automotive fire apparatus utilized for fire fighting and rescue operations.

NFPA 1914 Standard for Testing Fire Department Aerial Devices applies to the inspection and testing of all fire apparatus.

NFPA 1915 Standard for Fire Apparatus Preventive Maintenance Program defines the minimum requirements for establishing a preventive maintenance program for fire apparatus.

NFPA 1925 Standard on Marine Fire-Fighting Vessels provides the minimum requirements for marine firefighting vessels and also provides minimum maintenance and testing requirements.

NFPA 1931 Standard on Design of and Design Verification Tests for Fire Department Ground Ladders applies to the specific requirements for the design of and the design verification test for fire department ground ladders. This standard applies to all new ground ladders intended for use by fire department personnel for firefighting operations and training.

NFPA 1932 Standard on Use, Maintenance, and Service Testing of Fire Department Ground Ladders applies to the specific requirements for the use, maintenance, inspection, and service testing of fire department ground ladders.

NFPA 1936 Standard on Powered Rescue Tool Systems specifies the minimum requirements for the design, performance, testing, and certification of powered rescue tool systems including individual components of spreaders, rams, cutters, combination tools, power units, and power transmission cables, conduit, or hose.

NFPA 1961 Standard on Fire Hose applies to the testing of new fire hose, specified as attack hose, occupant use hose, forestry hose, and supply hose.

NFPA 1962 Standard for the Inspection, Care, and Use of Fire Hose, Couplings, and Nozzles and the Service Testing of Fire Hose applies to the care of all types of fire hose, coupling assemblies, and nozzles while in service, in use, and after use.

NFPA 1964 Standard for Spray Nozzles applies to portable adjustable pattern nozzles intended for general fire department use.

NFPA 1971 Standard on Protective Ensemble for Structural Fire Fighting specifies criteria and test methods for protective clothing designed to protect firefighters against environmental effects during structural firefighting (see Figure 2–8).

NFPA 1975 Standard on Station/Work Uniforms for Fire and Emergency Service specifies general requirements, performance requirements, and test methods for materials used in the construction of station/work uniforms to be worn by members of the fire service.

NFPA 1976 Standard on Protective Ensemble for Proximity Fire Fighting specifies design, performance, and test methods for protective clothing designed to provide limb/torso protection for firefighters.

NFPA 1977 Standard on Protective Clothing and Equipment for Wildland Fire Fighting specifies the minimum design, performance, testing, and certification requirements for protective clothing, helmets, gloves, and footwear that are designed to protect firefighters during wildland firefighting operations.

NFPA 1981 Standard on Open-Circuit Self-Contained Breathing Apparatus for Fire and Emergency Services specifies requirements for design, performance, testing, and certification of SCBA used in firefighting, rescue, and other hazardous situations.

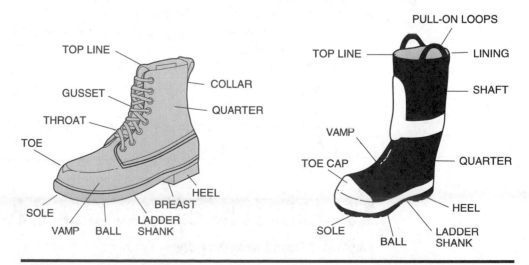

Figure 2–8   *Two types of* NFPA 1971-*compliant foot protection.* Courtesy USFA.

NFPA 1982 Standard on Personal Alert Safety Systems (PASS) specifies design, performance, and test methods for personal alert safety systems to be used by firefighters engaged in firefighting, rescue, and other hazardous duties.

NFPA 1983 Standard on Fire Service Life Safety Rope and System Components specifies performance, design, and test methods for life safety rope, harness, and hardware used by the fire service.

NFPA 1991 Standard on Vapor-Protective Ensembles for Hazardous Materials Emergencies applies to the requirements for vapor-protective suits and replacement components.

NFPA 1992 Standard on Liquid Splash-Protective Ensembles and Clothing for Hazardous Materials Emergencies applies to the requirements for liquid splash-protective suits and replacement components.

NFPA 1994 Standard on Protective Ensembles for Chemical/Biological Terrorism Incidents specifies the minimum requirements for the design, performance, testing, documentation, and certification of protective ensembles designed to protect fire and emergency services personnel from chemical/biological terrorism agents.

NFPA 1999 Standard on Protective Clothing for Emergency Medical Operations specifies minimum emergency medical clothing designed to protect emergency personnel or patients from exposure to liquidborne pathogens during emergency medical operations.

The following four standards and one recommended practice are not specifically referenced by *NFPA 1500* but also impact an organization's safety and health program and therefore deserve review.

NFPA 1041 Standard for Fire Service Instructor Professional Qualifications

NFPA 1402 Guide to Building Fire Service Training Centers

NFPA 1410 Standard on Training for Initial Emergency Scene Operations

NFPA 1451 Standard for a Fire Service Vehicle Operations Training Program

NFPA 1620 Recommended Practice for Pre-Incident Planning

OTHER RELATED STANDARDS AND REGULATIONS

American National Standards Institute

The American National Standards Institute (ANSI) is a national voluntary standard-making organization. Because ANSI uses the consensus process, its membership is comprised of representatives from many disciplines with expertise in the area in which the standard is being developed. These representatives may include professional, trade, technical, or consumer organizations, state and federal agencies, and individual companies.

Many of the standards developed by ANSI are in response to requirements in other regulations and standards. There are a number of ANSI standards relating to the performance of firefighter and emergency medical operations personal protective equipment such as helmets and goggles. Other approved consensus standards may become ANSI standards, for example the ANSI Safety and Health Standards Board also has approved some standards developed by the NFPA. In February 2003, ANSI established the Homeland Security Standards Panel (HSSP). This panel is to catalog, promote, accelerate, and coordinate the timely development of consensus standards within the national and international voluntary standards systems intended to meet identified homeland security needs and to communicate the existence of such standards appropriately to governmental units and the private sector.

The safety and health officer must research applicable ANSI standards during program development.

American Society of Testing and Materials

The American Society of Testing and Materials (ASTM) is a not-for-profit organization that develops standards for materials, products, and services. Using individuals and organizations from various disciplines the ASTM also is a consensus organization. While much of the ASTM's purpose is directed at consumer protection issues, some impact emergency service safety. The ASTM has generated several standards directed toward emergency medical system operations including training.

Environmental Protection Agency

The Environmental Protection Agency (EPA) has promulgated standard 40 CFR 311 that mirrors the OSHA Hazardous Waste Operations and Emergency Response document. The implication of this action is most important for organizations that do not have state OSHA plans. In these states the requirements of OSHA 1910.120 apply to public employers because of this EPA regulation. Therefore a public emergency service organization in a non-OSHA state that responds to, or has the potential to respond to, hazardous substance releases must comply with the OSHA requirements.

Centers for Disease Control and Prevention

Guidelines for the Prevention of Transmission of Human Immunodeficiency Virus and Hepatitis B Virus to Health-Care and Public-Safety Workers was released by the Centers for Disease Control and Prevention (CDC) in 1989. This document provides an overview of the modes of transmission of HIV in the workplace, an assessment of the risk of transmission under various assumptions, principles underlying the control of risk, and specific risk-control recommendations for employers and workers. The document also includes information on medical management of persons who have sustained an exposure in the workplace to these viruses. This document is very useful for a department developing the exposure control plan that is required by OSHA 1910.1030.

More recently the CDC has published information on anthrax and other biochemical agents and information for first responders when dealing with radiation emergencies. This information is essential as emergency services organizations become better prepared for terror attacks.

These are just two examples of the information available from the CDC. It has many documents available that deal with responder safety, particularly in the area of emergency medical operations.

National Institute for Occupational Safety and Health

The National Institute for Occupational Safety and Health (NIOSH) is part of the CDC and was created as part of the act that created OSHA. The primary mission of NIOSH is to reduce work-related injuries and illnesses. One function of NIOSH of interest to emergency services is the study of the Firefighter Fatality Investigation and Prevention program. This program is described in detail in Chapter 12.

From a regulation and standards standpoint, NIOSH publishes documents applicable to emergency service responders. Information is available on bloodborne infectious disease, personal protective equipment, and traumatic incident stress. NIOSH is working with the U.S. Army Soldier Biological and Chemical Command and the National Institute for Standards and Technology toward the development of appropriate standards and test procedures for all classes of respirators that will provide respiratory protection from chemical, biological, radiological, and nuclear (CBRN) agents inhalation hazards.

Ryan White Comprehensive AIDS Resources Emergency Act of 1990, Subtitle B

Another document that must be reviewed when managing infection control is the Ryan White Act. This act sets forth the requirements for employers to designate an infection control officer. Further, the act provides for the notification of employees by medical facilities after an exposure takes place. The safety program manager must be aware of the act as the notification procedures, the designated officer, and the followup procedures should become part of the department's program.

SUMMARY

Many regulations and standards from a number of sources relate to emergency service occupational safety and health. Regulations are laws and can be enforced by the agency that promulgated them. Standards, however, are developed by the consensus of experts in the subject area and only become enforceable if adopted by a state or local government. Although standards may not be officially adopted, they maybe used to establish standard of care or may be referenced by a regulation or used during litigation proceedings.

Many applicable safety and health regulations come from OSHA. These regulations cover public employees in the twenty-six states and territories that have state plans. In other states these regulations do not apply to public agencies. OSHA has regulations for response to hazardous materials incidents, confined space rescue, respiratory protection, and fire brigades. Within the Occupational Safety and Health Act is the General Duty Clause that has implications on operations not explicitly covered by other OSHA regulations.

Consensus standards provide a guide for safety and health programs. The NFPA publishes a number of specific standards relating to fire-fighter safety and health. These safety standards are found in the *NFPA 1500* series. A number of other NFPA standards are referenced within the *NFPA 1500* series and therefore become part of them by reference. Although a standard might not be formally adopted there can be standard of care implications.

A number of other regulations and standards also exist that apply to safety and health including guidelines from the Centers for Disease Control, and Prevention National Institute for Occupational Safety and Health, American Society of Testing and Materials, American National Standards Institute standards, the Environmental Protection Agency regulations, and the Ryan White Law. In order to develop a comprehensive safety and health program, these regulations and standards must be consulted. If they are law, they must be complied with, but if not law they can provide a framework for the safety program that is accepted and supported by national standards and trends.

Concluding Thought: Safety and health are ongoing considerations for which there are many laws and standards.

REVIEW QUESTIONS

1. The fire chief has asked you to sit on a committee to update standard operating procedures (SOPs). Consider each of the following SOP subjects and list what regulations or standards you would refer to as you formulated the updated SOPs.

   a. Confined space rescue

   b. Hazardous materials response and operations

   c. Infection control

   d. Incident management system

   e. Self-contained breathing apparatus maintenance and usage

   f. Operations at single-family house fires

2. Discuss the meaning of *standard of care.*

3. Which of the following states do not have state OSHA plans?

   a. Florida

   b. Maryland

   c. California

   d. Kentucky

4. Compare regulations with standards.

5. What OSHA regulation requires two in/two out at a structural fire?

   a. 1910.1030

   b. 1910.134

   c. 1910.156

   d. 1910.120

6. The guidelines published by the Centers for Disease Control and Prevention would help the safety program manager in complying with which OSHA regulation?

   a. 1910.120

   b. 1910.156

   c. 1910.134

   d. 1910.1030

7. In effect, 40 CFR 311 mirrors OSHA 1910.120 (True or False)

8. NIOSH is part of the Centers for Disease Control and Prevention and was created as part of the Act that created OSHA.

   a. True

   b. False

9. *NFPA 1500* is intended to apply to *all* fire departments including industry fire brigades.

   a. True

   b. False

10. You are the lead instructor in a training class that is going to conduct a live burn. Which of the following NFPA standards would be helpful in planning a safe drill?

    a. *1500*

    b. *1403*

    c. *1410*

    d. all of the above

ACTIVITIES

1. Review your department's purchasing process in terms of equipment and apparatus. Does your department follow NFPA recommendations?

2. Review your department's procedures for response to confined space and hazardous materials incidents. Are your procedures in compliance with those in the respective regulations?

3. Review *NFPA 1500*, then study your department. Determine where your department stands in terms of compliance.

Chapter 3

RISK MANAGEMENT

Learning Objectives

Upon completion of this chapter, you should be able to:

- Discuss the meaning of and the process for risk identification.
- Discuss the meaning of and the process for risk evaluation.
- Discuss the meaning of and the process for risk control.
- Apply risk control strategies to an injury problem.

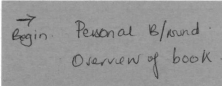

Begin. Personal B/round.
Overview of book.

CASE REVIEW

At approximately 0500 hours on the morning of June 28, 2002, a fire started in the kitchen of a condominium located on the southwest corner of the fifth floor of an eleven-story residential high-rise, built of fire resistive construction in Clearwater, Florida. The adult occupant of the unit attempted to fight the fire with portable fire extinguishers and an occupant-use fire hose from the building's standpipe system before notifying the fire department or activating the building's fire alarm system, which was monitored by a central station fire alarm service. The delayed alarm resulted in the death of two occupants of the building and injury to five firefighters, which included serious burns to one firefighter that required more than three months of recovery.

Challenges encountered by arriving firefighters included lack of knowledge of a working fire by the first engine company due to communication difficulties, a shutoff standpipe riser, an out-of-service hydrant on the south side of the building, and deviation from SOGs for alarms in high-rise structures. After addressing these initial difficulties, the operation recovered and the firefighters extinguished the fire with a combined exterior and interior attack. Three alarms were called because of the injured firefighters, the need to search and evacuate a large building with a wide occupant age range and evacuation capabilities, and the early problems encountered with insufficient fire flow to extinguish the fire. Risk management before, during, and after an incident is the basis for any safety and health program. Good risk management programs and practices are necessary to reduce illness, injuries, and deaths to emergency responders.

Source: From the United States Fire Administration Technical Report Series USFA-TR-148.

INTRODUCTION

■ **Note**
Risk management, required as part of *NFPA 1500* includes four components: risk identification, risk evaluation, risk control, and risk management monitoring.

Understanding risk management and implementation of a risk management plan are the basic building blocks of the occupational safety and health program. After the safety and health manager has reviewed the applicable standards, risk management must be undertaken. Risk management, required as part of *NFPA 1500* includes four components: risk identification, risk evaluation, risk control, and risk management monitoring. The first three of these concepts are presented in this chapter along with strategies to undertake to complete the process. The fourth component, risk management monitoring, is described in Chapter 11, Safety Program Evaluation.

RISK IDENTIFICATION

risks
the resultant outcome of exposure to a hazard

Risks could be defined as anything bad that could happen to an organization. A risk might be associated with a training exercise, an emergency medical incident, or simply a fall in the station. Identifying risks is complex, as in some cases you are trying to predict what could happen. Identification should include information from a number of sources including local and national resources.

Local Experience

The local experience can be reviewed in terms of historical incident data and data to help project future risks. Historical data can provide valuable information regarding risk and exposure based on the local history. For example, in analyzing the risk for firefighter injury from falls, the safety manager might review the equipment and procedures used to access aboveground locations. The severity and frequency of the risk is presented in the next section, Risk Evaluation. Another resource from local experience in the evaluation of the risk is to question the responders. Responders usually have the best knowledge of the area and have often identified certain potential risks within their response areas. Examples may be a simple risk, such as identifying an area of a highway that is prone to flood during rain storms or a problem with responding units meeting at certain intersections during a response. Both of these cases are clearly a risk in terms of vehicular accidents.

Historical and current data are useful for determining what has happened or what could happen today. Of greater difficulty is trying to predict what may happen in the future. Identification of future risk can be facilitated by contacting the local planning department, which can provide information on future road expansion, future industrial/commercial occupancies, and other changes that may occur in the community (see Figure 3–1). This information will not only help to identify future risks, but also provide time to plan for the risk. An example might be a town where the tallest building is one story. The town's fire department does an excellent job of risk identification and has developed a risk management plan for the risks that have been identified. However, the city is planning to build a

> **■ Note**
> Another resource from local experience in the evaluation of the risk is to question the responders.

Figure 3–1 *Changes in construction types occurring in the community should be evaluated as part of risk identification.*

multistory senior citizen residential building. This planned change will significantly change the department's risk management plan as now risks include those that are common in high-rise structures.

Identification of Trends

Maintaining currency with emerging trends can also be useful in risk identification. By reading trade journals, searching the Internet, and attending conferences and seminars, the safety manager can see emerging trends. For example, the trend for fire departments to become involved in emergency medical response clearly created additional risks; the move toward hazardous material response and control created additional the risks. The trend of building houses and apartments with lightweight truss roof construction increased the risks during interior firefighting. The move toward placing supplemental restraints, airbags, in cars increases the risk to responders to vehicle accidents.

Trend analysis can help to identify future as well as current risk. The trend of reducing staffing and increasing services provided by emergency service organizations is certain to create additional risks. Some EMS systems now provide inoculations. Should your department include this involvement in the future, getting stuck with a needle and the concern for cross contamination may be an identified future risk.

Safety Audit

The safety audit can be another way of identifying risk within an organization. The audit should be based on some type of checklist in which a reviewer goes through the organization and checks for compliance. Commonly in emergency services safety programs, the *NFPA 1500* standard is used as a benchmark. Deficiencies are noted and any associated risks identified. For example, the *NFPA 1500* standard requires that all newly purchased protective gloves comply with the current NFPA standard on fire service gloves. If the safety audit reveals that half of the department's gloves have not been purchased under the new standard, the changes in the new standard should be reviewed against the quality of the existing gloves and the potential risk of hand injury should be identified. Safety audits can be performed in-house by safety staff or contracted to an outside safety consultant.

Reviewing Previous Injury Experience

Clearly one of the best ways to identify risk in terms of safety and health is to review past statistics. Within an organization this review can come from injury reports, workers' compensation history, or if required, the OSHA illness/injury logs. As described in Chapter 1, it is also important to compare the local statistics to those nationwide and to form an understanding of why they may differ. The review of previous injuries complements the risk identification process basically because, if it happened and it caused an injury, it must be a risk. The one downfall in the process is that you are looking at a history and just because something has happened, it may never happen again, or just because something has not happened, does not mean it will not happen.

Once risks are identified through the four processes described previously, it is useful to categorize the risk. Categories might include responding, in station, training, fire or EMS emergency, natural or man-made disasters, and hazardous materials incidents. Categories help the safety manager to later sort the risks.

■ **Note**

Categories help the safety manager to later sort the risks.

RISK EVALUATION

Once the risks are identified, they should be recorded for future reference as the next step in the process is to evaluate the risks. For risk evaluation, we introduce two risk management terms, *frequency* and *severity*. Risk evaluation is done from these two perspectives.

Frequency

frequency
how often a risk occurs or is expected to occur

Frequency is simply how often a particular risk is likely to occur, be it a fall or an auto collision. The safety manager must evaluate how often a risk will occur in order to better assign the risk a priority later. Unfortunately there is no hard-and-fast rule for what an acceptable frequency is. It is a somewhat subjective measure that is very dependent on local conditions and the person making the evaluation. For this reason, numerical measures are not assigned to frequency, but instead just a high, medium, or low.

■ **Note**

The safety manager must evaluate how often a risk will occur in order to better assign the risk a priority later.

For example, suppose our department, on any given day, has 100 emergency vehicles on the highways responding to incidents. Using historical data we determine that there are fifty traffic accidents per year, about one a week, involving the emergency vehicles. Is this frequency high, medium, or low? What about a department with twenty-five emergency vehicles and the same accident rate? Sounds high? So maybe in the smaller department the risk would get a high frequency rating. However, what if the first department had just spent a year training drivers in modified responses to nonemergency alarms and installed intersection control lights to prevent accidents? This fifty then would be considered a high frequency for them as well.

One way to determine if a frequency is high, medium, or low is to benchmark the local experience to that nationwide or to that of similar size organizations. For example, if the safety manager for the smaller department selected similar-sized departments with similar demographics and a similar number of vehicles on the road and found out the average number of accidents per year was 100, then the safety manager may rate this frequency as low.

Severity

severity
how severe the result
is when a risk occurs

Severity is a measurement of how great the loss or the consequences of the loss will be. There are two variables in the measurement of severity, but again some is in the judgment of the person measuring. As with frequency, a measurement of high, medium, and low is used as opposed to a numerical rating. The following factors must also be considered when determining the severity element:

Costs Costs of a risk are measurable, however, cost can be both direct and indirect. Direct cost might include the costs of medical treatment, the overtime paid to cover a vacancy on a crew, or the cost of replacing equipment. Indirect costs are more difficult to measure, such as the loss of productivity, the loss of using the equipment, stress-related concern of coworker, morale, and possibly the cost of replacing the employee.

Organizational Impact How would the risk impact the organization? For example, the collision of a department's only ambulance would put that department out of the EMS transport business until a replacement could be put into service. Smaller departments could be crippled by an incident in which multiple employees are injured, such as a chemical exposure. In a 1947 Texas City, Texas, explosion, a shipboard fire and explosion involving ammonium nitrate killed almost all of the fire department personnel. Also included in organizational impact is the time it would take to recover, which might be the time it takes to rehire or retrain personnel or replace an emergency vehicle.

Frequency and Severity Together

When the program manager combines frequency and severity, the result can help to determine priorities to which risk control measures can be applied. It is common and necessary to prioritize the risks, as seldom are time, money, and other required resources sufficient to control all of them. In prioritizing, it is important to remember that these two measures are somewhat subjective and very organizationally specific. However they can be quite useful.

■ **Note**
In prioritizing, it is
important to remember
that these two
measures are some-
what subjective and
very organizationally
specific.

Let us return to our example of the town that is about to get a high-rise building for senior citizens. For simplicity, let us say that we have determined three risks associated with this change. The risks identified are: collisions while responding, trapped during firefighting, and sprains and strains associated with lifting during medical responses. A quick table can be put together (see Table 3–1) and an analysis made.

These ratings may have been given after significant study. For example, the building is to be built across from the EMS station, therefore the risk of a vehicle accident en route would be low. However, if an accident were to occur, the severity would be high because the department only has one ambulance. The building

Table 3–1 *Frequency and severity of identified risks.*

Risk	Frequency	Severity
Collisions while responding	L	H
Trapped during firefighting	L	H
Sprains and strains associated with lifting during medical responses	H	M

is being built to current fire codes and has a state-of-the-art private fire protection and detection system. Although the severity of becoming trapped in a high-rise is high, the potential for this to occur in this building is low. Finally, the safety manager in this case called a fire/EMS department in another town that has a similar building. That department reported that it responds to a minimum of one EMS incident per day at their senior high-rise and usually transports someone. The safety manager further researched his department statistics and found that the department had a higher-than-average back injury problem. Therefore, in the high-rise case, the priority risk should be the prevention of back injuries. This is not to say that the other risks are not important, only to provide a guideline for prioritization.

RISK CONTROL

risk control

a common approach to risk management in which measures and processes are implemented to help control the number and the severity of losses or consequences of risk to the organization

Once the risk identification and evaluation process has been completed, the safety manager must develop ways to control the risk. **Risk control** can be divided into three broad categories: risk avoidance, risk reduction, or risk transfer.

Risk Avoidance

To avoid risk is simply not to do the task with which the risk is associated, but this option is not usually available to the emergency response profession. By the nature of the profession, we are called when things have already gone wrong. Many times it is not possible to avoid the risk. For example, the risk of steam burns from fighting an interior fire could be completely eliminated if the fire department chose not to enter burning dwellings. If the risk of exposure to infectious diseases were a priority, we could avoid the risk by not responding to calls involving blood products. But it would be difficult to justify our existence if we chose this avoidance route and did not respond to these types of incidents.

However, some risks can be dealt with successfully by avoiding the risk. For example, a risk associated with starting an intravenous line (IV) is getting stuck with the needle when trying to resheath it. So to avoid the risk, we develop a policy that sets the procedure for dirty needles to be put directly into a sharps box

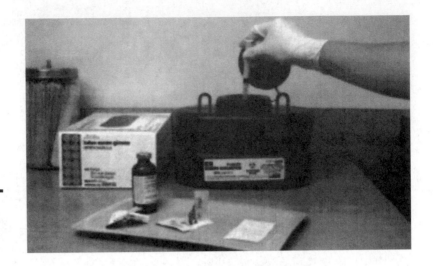

Figure 3–2 *Sharps boxes are used to avoid needle-stick injuries.*

without resheathing them (see Figure 3–2). A more costly endeavor might be the risk associated with backing the fire apparatus into the bays. To avoid the risk, the department could design future stations with drive-through bays.

Risk Reduction

If the risk cannot be avoided, then the next best effort that can be put forth is to reduce the risk or minimize the chance that it will occur. There are many ways to develop risk reduction measures and there may be several ways to control a particular risk. Any development of risk reduction measures must involve input from the employees who are affected. Risk reduction measures that are not or cannot be followed are useless.

> Apply to Eg ATE setting or T3:2.

Interrupting the Accident Sequence One method of developing risk reduction is to view the injury process as a series of interruptible events. The accident sequence can be described in terms of five factors: the social environment, human factors, unsafe acts or conditions, accident, and the injury. The principle behind this thinking is that any interruption in one of the factors will prevent the injury or risk from occurring. Heinrich[1] makes this theory analogous with dominos standing on end. If one domino is pushed, they all will fall. But if the center domino is removed, the falling will stop there.

In using this concept to generate risk reduction measures, the safety manager can look at past and predicted future risks and break the sequence down into events. Then these events can be classified according to the five factors in the sequence and control measures developed.

■ Note

Any development of risk reduction measures must involve input from the employees who are affected.

■ Note

The principle behind this thinking is that any interruption in one of the factors will prevent the injury or risk from occurring.

---

[1]Heinrich, H. W., *Industrial Accident Prevention: A Scientific Approach* (4th ed), New York: McGraw-Hill, 1959.

Table 3–2 *Classification of events leading to back injury.*

Event	Social Environment	Human Factor	Unsafe Act or Condition	Accident	Injury
Location of equipment on apparatus			X		
Fitness of the firefighter		X			
Not wanting to ask for help	X				
Muscle strain					X
Overexertion				X	

Haddon matrix
a 4 × 3 matrix used to help analyze injuries and accidents in an attempt to determine processes designed to reduce them

matrix
a chart used to categorize actions or events for analysis

■ Note
Human factors are related to the employee or person involved in the injury. Human factors can contribute to the injury from three perspectives: poor attitude, inadequate training and education, or lack of fitness for the task.

Let us take, for example, back injuries that occur from lifting heavy equipment from overhead compartments on the apparatus. The safety committee breaks down the events that lead up to the injury, then each event is classified. Table 3–2, which is illustrative and not exhaustive, shows the result of this process.

Once the group has brainstormed all events and classified them, control measures can be developed. In this case, procedures could be written that require two persons to lift heavy equipment, apparatus can be designed for heavy equipment to be stored low, training could be undertaken to improve back strength and develop proper lifting techniques. Although this example is simple, the process can be used on much more complex problems.

The Haddon Matrix Another method of developing risk reduction strategies can be borrowed from a long-time injury prevention expert and developer of the **Haddon matrix**, Dr. William Haddon. Haddon based the **matrix** on the process of analyzing injuries. The matrix implies that there are four components of the injury process and three time intervals. The four components are the human factors, the energy vector, the physical environment, and the social environment. Human factors are related to the employee or person involved in the injury. Human factors can contribute to the injury from three perspectives: poor attitude, inadequate training and education, or lack of fitness for the task. The energy vector is what causes the injury. For a burn it would be the heat of fire or steam; for a back strain it might be the patient or piece of equipment, and for a laceration it could be the glass from a window or jagged metal on a damaged automobile. The physical environment is where the risk is likely to occur. This factor includes weather conditions, seating area on fire apparatus, a natural disaster, or placement of equipment on the fire apparatus. The fourth component is the social environment, which involves the organization's climate, values and beliefs of employees, and peer pressure.

The time interval of preevent is studied to try to prevent the injury or risk from occurring in the first place. The event interval focuses on the risk at the time it is occurring; the strategies are aimed at reducing the exposure or minimizing the risk. Postevent is after the event has occurred and focuses on early notice that an event has occurred, proper care if necessary, and proper rehabilitation and recovery.

Risk reduction measures can be generated using this matrix. Although some of the control measures that could be generated may not be practical, many will be. A good way to use this tool is to select priority risk and have a group, often the safety committee, brainstorm control methods that would fit into each of the twelve blocks in the matrix. The committee or group can then go back and eliminate those that are not practical and further research those that are.

A result of this process is shown in Table 3–3, which depicts a partially completed matrix attempting to minimize burn injures after a brainstorming session by department officers and the safety program manager. This table is illustrative, not exhaustive. Local conditions and procedures may not allow for many of the measures to be implemented.

The Haddon thinking goes yet another step in the risk reduction strategy process. Once the measures that are practical are selected, the measures are further divided as to whether they are passive or active, and voluntary or mandatory. Passive here means that you do not have to do anything special for the control

Table 3–3 *Example of a completed matrix.*

	Human	Energy	Physical Environment	Social Environment
Preevent	Train firefighters and company officers to recognize flashover conditions Have incident commanders make a risk analysis before committing to interior attacks on building not worth saving	Do not go in	Separate from cause Use proper hose stream and adequate gallons per minute Follow good tactics including ventilation	Education of public to be safer with fire Less peer pressure about wearing personal protective equipment (PPE)
Event	Train on survival techniques and rapid escape Have a rapid intervention team at the ready Wear a self-on PASS device	Wear full PPE	Have an escape route	Less aggressive attacks
Postevent	Training in burn care	Make others aware of fireground or interior conditions	EMS at scene for rapid care and transport	Proper follow-up care and rehabilitation

measure to occur; active means that you must physically do something. Voluntary measures allow people to choose, whereas mandatory measures are mandated by laws, rules, or regulations. It is evident that the most effective measures are those that are mandatory yet passive. Using PASS devices is common throughout the fire service. When they first came out, some individuals purchased them for themselves. This risk reduction method would have been considered voluntary and active. It is hard to predict how effective it was. Now PASS devices are a mandatory part of safety standards. However, until recently they required an active process, namely turning them on. Again, effectiveness was in question, because many were not activated while fighting a fire. Recent improvements have the PASS device turn on with the breathing apparatus, hence a passive process and an increase in effectiveness.

Some of the risk reduction measures from the burn example from Table 3–3 are further categorized in Table 3–4. This table is also for illustrative purposes and is not to be considered exhaustive. Note that some control measures may fit into more than one category.

Using the 4 × 3 and 2 × 2 matrix can be a valuable tool to assist the safety program manager in developing risk control methods. It is easy to understand and provides the categorization necessary to formulate effective strategies.

Risk Transfer

risk transfer
the process of letting someone else assume the risk; for example, buying auto insurance transfers the consequences of an accident to the insurance company

Risk transfer in emergency services is most commonly associated with the purchase of insurance policies. The transfer of risk does nothing to prevent the risk and only limits the amount of financial exposure for the organization. The transfer of risk would do very little in terms of the time it takes to hire a replacement firefighter for one that was injured and retired on a disability pension. Risk transfer is closely related to the safety and health program, as the amount that is paid for insurance is based on previous experience, including injuries, accidents, and other risks.

Table 3–4   *A completed matrix comparing voluntary and mandatory passive and active measures.*

	Voluntary	Mandatory
Passive	EMS at scene for rapid care and transport	Wear a self-on PASS device
Active	Use proper hose stream and adequate gallons per minute	Wear full PPE
	Have an escape route	Use proper hose stream and adequate gallons per minute
		Wear full PPE

Risk transfer through insurance is common. Insurance carriers are interested in the operation of the department from a risk management standpoint, not only to reduce claims, but in today's competitive environment many organizations bid their insurance coverages. The insurance company, therefore, also has an interest in keeping premiums low. Expect an insurance company to look favorably on things like driver training programs, SOPs, drug-free workplace policies, and safety committees, to name a few.

However, the issue of risk transfer through insurance is a complex one, well beyond the scope of this book so only a brief explanation of different types of insurance is provided.

Workers' Compensation Insurance Workers' compensation insurance is the insurance that covers employees injured while on the job. Each state has different requirements for workers' compensation so there is a great deal of variation in requirements and benefits. Generally, workers' compensation provides for the payment of medical expenses required after an illness or injury suffered on the job. Workers' compensation also pays a wage to the employee to cover time off following a job-related illness or injury. Because the wage that workers' compensation provides is only a portion of the employee's normal salary, there is incentive to get back to work. Some employment agreements provide for 100% of salary for a specified period of time after an injury or illness suffered on the job. In this case, the employer is required to make up the difference and may carry a secondary insurance policy.

Management Liability Sometimes referred to as errors and omissions, management liability insurance is designed to cover the actions, or lack thereof, of employees while performing duties on behalf of the employer.

General Liability General liability insurance is designed to protect the organization from a property loss. Although coverage varies, protection against theft, fire, storms, and such are usually provided.

Vehicle Insurance Vehicle insurance covers the organization's vehicles from damage or theft. Vehicle insurance policies also protect the insured from liability caused by the vehicle or its operator. Vehicle insurance generally does not cover an on-duty employee from an injury standpoint. Required medical treatment after a collision is covered by workers' compensation.

Regardless of an organization's safety record or safety and health plan, risk transfer is necessary. The cost of a single incident, such as a vehicle collision, could easily bankrupt a department without insurance. If injuries occurred, there would be property loss, medical expenses and lost time from injured employees, and liability for injuries to occupants of the other vehicles involved.

SUMMARY

Risk management is a three-step process that requires risk identification, risk evaluation, and risk control. Risk identification, the first step, allows the safety manager to identify risks both from a historical basis and to predict those that are likely to occur in the future. Risk identification is done by analyzing local experience, identifying trends, conducting a safety audit, and reviewing the previous injury experience.

Risk evaluation is a measure of both frequency (how often a risk could occur) and severity (how bad it will be if it does). Together the evaluation of risk helps the safety manager set priorities for risk control programs

Risk control is the third step in the risk management process. Risk control can be accomplished by avoiding the risk, reducing the risk, or transferring the risk. In emergency services, risk avoidance has narrow application but in some cases can be useful. Risk reduction, however, can be used aggressively. The tools used to develop risk reduction strategies are interruption of the accident sequence and the Haddon matrix. Interruption of the accident sequence requires the classification of the accident events in order to develop measures to interrupt the process at some point. The Haddon matrix is a 4×3 matrix that considers four factors of three time intervals. The Haddon process further classifies the prevention measures into a 2×2 matrix as being active or passive and voluntary or mandatory.

Risk transfer is the process of transferring the risk to someone else. For the purpose of this text, the transfer would be to an insurance company. There are several types of insurance such as workers' compensation, vehicle damage and liability, property damage and loss, and errors and omission. Requirements for insurance are varied and beyond the scope of this text. Readers are encouraged to check local and state laws for insurance requirements.

Concluding Thought: Safety and health program management is not only subjective. It has objective components that are the foundation for local decision making.

REVIEW QUESTIONS

1. List, in order, the four steps in the risk management process required in *NFPA 1500*.

2. What four areas can be examined when performing the risk identification process?

3. Explain the difference in the terms *frequency* and *severity*.

4. Give an example of risk avoidance.

5. List and explain, in general terms, the four kinds of insurance coverage.

6. What are the twelve components of the Haddon matrix?

7. Which of the following would be most effective as a risk reduction measure?

   A. Voluntary/Active
   B. Mandatory/Passive
   C. Mandatory/Active
   D. Voluntary/Passive

8. What are the four classifications in the accident sequence, according to the Haddon matrix?

9. Give two examples of human factors in the Haddon matrix.

10. Define the social environment.

ACTIVITIES

1. Select an injury class from your department's statistics, for example, back injuries while advancing hose lines. Apply the Haddon matrix and develop some risk control measures.

2. Using the same risk as in Activity 1, apply the interrupting-the-accident sequence process. Compare the similarities and differences. Which process produced better results?

Chapter 4

PREINCIDENT SAFETY

Learning Objectives

Upon completion of this chapter, you should be able to:

- Describe safety considerations in the emergency response station.
- Explain safety considerations as they apply to the emergency response vehicle.
- List the components of an effective response safety plan.
- Describe the components of a preincident planning process.
- List the information that should be provided by the preincident plan.
- Describe the considerations for safety while training.
- Define the components of a wellness/fitness plan.
- Describe the considerations for interagency coordination as it applies to health and safety.

CASE REVIEW

In August 1996, the IAFC Fire-Rescue International was being held in Kansas City, Missouri. On Saturday, August 24, the annual Combat Challenge was scheduled. On that day, members of the IAFF Local 42 and other IAFF locals protested the event with informational pickets. The picketing was intended to allow firefighters to exercise their right to express grievances, primarily about the way the Combat Challenge allegedly has been used in a noncompetitive, job-threatening manner in some departments. The picket line grew to nearly 200 supporters. Following reports of intimidation by picketers, assaults, and damage to competitors' vehicles, the IAFC canceled the competition.

After this incident, IAFC President Chief R. David Paulison was quoted as saying "We must now take the opportunity for management and labor to explore the issues and try to reach an agreeable approach to firefighter physical fitness." And that is exactly what happened. Fire chiefs and union leaders from ten fire departments from across the United States and Canada were selected to sit on this committee and develop a comprehensive physical fitness program.

One year later, the IAFF/IAFC joint physical fitness committee presented a program at both the IAFC's Fire Rescue International and at the IAFF's Redmond Symposium. The program is a complete nonpunitive program that includes requirements for medical examinations, physical and mental fitness, rehabilitation, and wellness.

The complete program and training package are available from the IAFC or through IAFF locals.

INTRODUCTION

During the risk identification process, a number of risks can be identified that occur before an incident does. These risks focus on issues relating to the workplace, the equipment or vehicle being used, and the human resource that is using it or working in the workplace. In this chapter, the focus is on risks that occur prior to an incident or at the time of an incident. Many of these risks are the easiest to manage, because, with few exceptions, the risk occurs under a controlled situation or environment.

Preincident health and safety can be divided into seven categories:

1. Station safety
2. Apparatus safety
3. Response safety
4. Preincident planning
5. Safety while training
6. Fitness/wellness
7. Interagency cooperation in safety-related matters

STATION CONSIDERATIONS

Consideration for safety at the station is grouped into two broad categories: design and ongoing operations. The design categories deal with fire station design prior to construction or at the time of renovation, for example provisions for an internal fire protection or exhaust emission system (see Figure 4–1). Ongoing operations look at safety and health issues from a day-to-day operational basis, such as wiping up wet spots to prevent falls or station safety inspections.

Design

When it comes to facility safety, public safety agencies should serve as models in the community. Therefore, during the design phase of the fire/EMS station, safety must be an important design consideration. The station must be built to comply with all applicable standards and codes. Chapter 9 of *NFPA 1500* requires compliance with all applicable health, safety, building, and fire codes for fire stations (see Figure 4–2). Stations must be designed with private fire detection and protection systems such as sprinkler systems, smoke and heat detectors, and carbon monoxide detectors.

Once the design is in compliance with local codes and standards, it is important to look at the physical design. Proper routes of travel must be provided from living areas to the apparatus bays. These routes of travel must be direct, without obstruction, and lighted for night operations. Sleeping areas must be designed so that after dark, low-level lighting is provided to prevent trip and fall injuries. If the station is multistory and chutes or poles are used, there must be protection from

Figure 4–1 *There are many different designs for fire stations. Each should consider responder safety during day-to-day operations.* (Courtesy Palm Harbor Fire Rescue.)

Figure 4–2 *Fire stations should be designed to meet all local building and fire codes.*

employees falling down them by accident, and new employees must be trained on the proper use of these devices. The bays should be provided with proper drainage to limit standing water, and walk areas should be painted using a nonslip paint. Bathroom floor and shower areas should be nonslip surfaces as well.

Storage areas for protective clothing must allow for ventilation and must not be in living areas. Unless contracted out, a laundry facility may be provided. By OSHA regulation, there must be an area for the cleanup of materials contaminated with bloodborne pathogens.

The bays should have some form of exhaust emission system. Exhaust systems may be the type that attach to the vehicle's exhaust by a tube and are vented to the outside. A second system vents the entire apparatus bay area, using roof or wall-mounted exhaust fans and louvers in the bay doors. This system is useful in colder climates in which small equipment, such as rescue tools and saws, must be run inside the bays. Yet another exhaust elimination system is not part of the station, but attaches and remains on the emergency vehicle exhaust.

The station design process should include input from the safety committee and the crews themselves. The crew will be able to pinpoint safety issues in the design of a building (see Table 4–1).

It clearly is much easier and less costly to design the previously mentioned features into a new building prior to construction. However, most of the facilities that the safety manager deals with today have long been constructed. Some are more than 100 years old. The safety manager, in conjunction with fire department management and the safety committee, must prioritize improvements and try to retrofit the station with the necessary safety features over several budget years. Modifications that are required by regulation may have to be made early on.

■ **Note**
By OSHA regulation, there must be an area for the cleanup of materials contaminated with bloodborne pathogens.

■ **Note**
The station design process should include input from the safety committee and the crews themselves.

Table 4–1 *Example station design checklist.*

	Yes	No
Applicable Code Compliance		
Health		
Safety		
Building/life Safety		
Fire Prevention		
Physical Design		
Routes of travel—direct, marked, free of obstruction, lighted		
Sleeping areas—low-level lighting provided		
Bathroom areas—nonslip surfaces		
Poles or chutes protected		
Apparatus bays—Drainage		
Apparatus bays—Lighting		
Apparatus bays—Exhaust emission system		
Apparatus bays—Nonslip surface		
Apparatus bays—Doors: Self-reverse safety system		
Apparatus bay access—Drive through preferred		
Biohazard cleanup room/area		
PPE storage room with ventilation—not in living areas		
Station access—proper design—ample parking		
Proper and ample storage areas provided		
Safety and Security		
Ability to secure station		
Fire alarm system with smoke detectors		
Sprinkler system		
Carbon monoxide detectors		
Provisions to protect employees from extreme weather		

Sufficient parking should also be provided, particularly for volunteer departments where members respond to the station for the apparatus. Responders should be able to park where they can exit their vehicles and access the station safely.

Ongoing Operations

Of additional concern regarding station safety, and probably more important, are the procedures and safety measures relating to the day-to-day operations of the station and the inhabitants, the responders, and the public that might visit. The station is generally a controlled environment, so injuries that occur at the station could often easily have been prevented.

■ **Note**

The fire station should have a complete fire inspection at least annually in accordance with local code.

First and foremost, the fire station should have a complete fire inspection at least annually in accordance with local code. Fire protection systems should be certified as recommended or required in the code. Smoke and carbon monoxide detectors should be checked according to manufacturer's recommendation. Generally, the on-duty crew can perform these inspections, sometimes with assistance from the fire prevention division and using standard forms (see Figure 4–3). Local OSHA or state departments of labor will usually help a department identify safety problems in the station, as will many insurance carriers. One insurance company even provides inspection forms.

Aside from the fire safety concern, station inspections that look for hazards and safety issues should be conducted daily. Part of the shift change process should require the on-coming supervisor to walk the station area and look for safety or health problems. These might include wet spots on floor, the proper operation of an emergency bay door stop, or burnt out lights in a bedroom. Further department procedures should forbid unsafe acts such as creating tripping hazards with debris or electrical cords. The procedures can also be proactive, such as using Wet Floor signs after mopping a floor. It is interesting that on the fireground a firefighter will climb 15 feet to a second-floor window for an operation and using a leg lock, lock into the ladder. But in the station that same firefighter will climb 25 feet on a stepladder to change a light bulb and not consider tying off. Procedures and training should prohibit this type of behavior. Some examples of station safety policies are provided in Text Box 4–1 and Text Box 4–2.

Text Box 4–1   *Example policy.*

FACILITY SAFETY

Purpose

Fire stations, like all structures, are subject to fire and safety hazards. This standard is designed to keep fire stations as safe as possible to protect the lives and health of the personnel within.

Definition

Facility safety is the result of regular inspections for various hazards and conscientious personnel who are always on the lookout for dangerous situations and acts.

(continues on page 68)

STATION INSPECTION REPORT

Station _____   Shift _____   Date _____

Captain _____   Battalion Chief _____   Division Chief _____

✔—O.K.
✘—See Remarks

1. Personnel

Uniform		
Protective Clothing		
Grooming & Cleanliness		
Driver's License		
I.D. Card		
E.C. Card		

5. Safety

Overhead Storage	
Caution Signs	
Safety Bulletin Board	
Tool Storage	
Combustible Storage	
Insecticide Chemical Stor.	
SCBA Tests	
Extinguisher(s)	
Smoke Detector(s)	
RADEF Equipment	

6. Exterior

Paint	
Doors	
Sidewalks	
Ramps	
Parking Area	
Yard	
Hose Tower	
Gas Pump	
Flag Pole	

2. Office

Files	
Maps	
Log Book	
Shift Change Log	
Desk	
Bulletin Board	
Tourist Information	

3. Engine Hose

Floor	
Work Bench	
Hose Rack	
Storage	
Turnouts	
Apparatus	

7. Interior

	Office	Kitchen	Day Room	Dormitory	Rest Room	Heater Room	
General Appearance							
Cleanliness							
Floors							
Windows							
Lights							
Furniture							
Storage Area(s)							
Woodwork							
Walls							
Ceiling							

4. Programs

Hazard Reduction	
In-Service Inspections	
Pre-Fire Plans	
Water Systems	

Remarks _____

_____

_____

_____

_____

Figure 4–3 *An example of a station inspection record.*

Objectives

- To schedule regular monthly and annual inspections of all fire stations.
- To provide for the immediate correction, or to be corrected as soon as possible, of all hazards and code violations found during said inspections.
- For all personnel to become safety conscious about their individual stations.

Station Inspection

Annual inspections. An annual inspection of each station will be done by a company in-service inspector (minimum) and will be accompanied by the shift captain or his designee. The inspection will be done on the first Thursday of September using the standard Safety Survey Form. Copies of the form will be given to the Fire Prevention Bureau and to the shift captain or his designee. Corrections will be made as soon as possible.

Monthly Inspections. A monthly walk-through inspection will be done on the first Thursday of each month using forms provided. The intent of this inspection is to identify unsafe conditions, check that exit lights and smoke detectors work properly, and that extinguishers and the Ansul System are properly charged. Department fire sprinkler forms shall also be completed and attached to inspection form. Hazards or violations will be reported in writing to the Fire Prevention Bureau and the shift captain or his designee. Corrections will be made as soon as possible. The shift captain or his designee will accompany the inspector on these inspections.

Text Box 4–2 *Example safety SOP.*

Office Safety (All Personnel)

A. Do not connect multiple electrical devices to a single outlet.

B. Do not use extension or other power cords that are cut, frayed, or damaged.

C. Close file and desk drawers when unattended. Do not open more than one drawer at a time and close it when done.

D. Put heavy files in bottom drawers of file cabinets to prevent cabinets from tipping over.

E. Do not tilt your chair back on two legs.

F. Do not use chairs, boxes, or other items as improvised climbing devices.

G. Turn off the machine and disconnect electrical power before attempting to adjust or clear electrical office equipment.

H. Do not remove, bypass, or tamper with electrical equipment fuses, switches, or safeguards.

I. Do not place your fingers in or near the feed of a paper shredder. Verify guards are in place and working prior to use.

Station Safety (All Personnel)

A. Mop/clean up oil, hydraulic fluid, water, or grease from apparatus floors and accesses, immediately, upon detection.

B. Safety goggles or other provided eye protection shall be worn when operating power equipment.

C. Do not run electrical and other cords across doorways, aisles, or between desks or create trip hazards.

D. Pick up all foreign objects, such as pencils, bolts, and similar objects, from floors to prevent slipping.

E. Clean up all spills immediately, especially wet spots around drink and coffee machines, in bathrooms, kitchen, and hallways.

F. Use stepladders only on a firm level base, opened to the full position with spreaders locked.

G. Do not leave tools in work areas, walkways, or stairways.

H. Do not walk across wet or oily areas. Avoid the hazard by walking around it.

I. Avoid walking across areas that have been freshly mopped.

J. Do not run on stairs or steps or take two at a time.

K. Use handrails.

L. Do not block your own view by carrying large objects.

M. Do not jump from trucks, platforms, scaffolds, ladders, roofs, or other elevated places, regardless of height.

N. Do not take shortcuts. Use aisles, walkways, or sidewalks.

O. Do not jump from, to, or between elevated areas.

APPARATUS SAFETY

Safety involving apparatus also has some preincident considerations. Like stations, apparatus considerations can also be categorized into design and ongoing operations. Design concerns are those that are analyzed and controlled at the time of purchase. Ongoing concerns are those that affect day-to-day operations. One major issue associated with apparatus is driving and responding to emergencies; that issue is addressed in the next section.

Design

Emergency service vehicles should be designed to meet applicable standards. There are NFPA standards that govern requirements for fire apparatus (see Table 4–2), many of which are based on firefighter safety needs. There are also

Table 4–2   *NFPA standards impacting apparatus design and ongoing operations.*

Standard Number	Title
NFPA 1901	*Standard for Automotive Fire Apparatus*
NFPA 1906	*Standard for Wildland Fire Apparatus*
NFPA 1925	*Standard on Marine Fire-Fighting Vessels*
NFPA 1912	*Standard for Fire Apparatus Refurbishing*
NFPA 1915	*Standard for Fire Apparatus Preventive Maintenance Program*
NFPA 1071	*Standard for Emergency Vehicle Technician Professional Qualifications*
NFPA 1911	*Standard for Service Tests of Fire Pump Systems on Fire Apparatus*
NFPA 1914	*Standard for Testing Fire Department Aerial Devices*
NFPA 1581	*Standard on Fire Department Infection Control Program*

requirements placed on the vehicles by the Department of Transportation, because the vehicle, in most cases, will travel on public roads. Ambulances also have federal regulations from the Department of Transportation (see Figure 4–4).

Although designing and meeting the minimum requirements stated in the standards provides some level of safety, the safety manager and safety committee

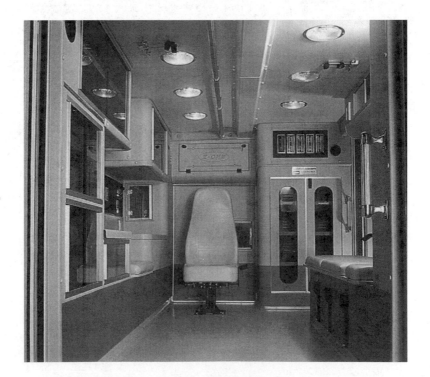

Figure 4–4

Ambulance design is governed by the department of transportation in the KKK specifications. Here is an example of interior layout. (Courtesy of E-One, Inc., Ocala, FL.)

should have direct involvement in the design process. A risk identification and analysis should be part of the design phase. For example the placement of equipment affects compartmentalization requirements and location. How will ladders be stored? If attached to the side of the apparatus, will they be too high to be safely removed? How about hose bed height for both supply line and preconnects? Can they be safely deployed and, just as important, safely repacked? The standard allows for the first step into the cab to be 24 inches from the ground. Is this low enough for a department that has identified ankle injuries caused by getting out of the cab as a risk? Or could another step be installed that might lower the step level to 18 inches? What about the design of the patient compartment of the ambulance? Again, the standards provide some level of safety, but the unit should be designed around local needs and conditions (see Figure 4–5).

Remember, manufacturers of emergency vehicles are willing to do just about anything possible to design a vehicle to meet the needs of their customer. However, many times modifications cost money and may be deleted in favor of other requirements for the apparatus. The safety manager's responsibility is to perform a thorough risk analysis and support each and every safety consideration.

Ongoing Operational Concerns

Most departments keep emergency vehicles for a number of years. A lot can be said for a vehicle that is designed properly, but most of the safety issues relating to apparatus are of the day-to-day operational nature.

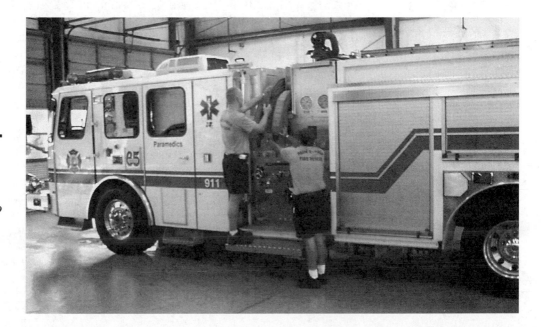

Figure 4–5
Apparatus must be designed to meet standards and regulations and also address local need and conditions. Note pull-out step that creates a safe platform for the members to reload the hose.

Figure 4–6 *A preventive maintenance program is integral to having a safe response vehicle.*

preventive maintenance (PM) program

an ongoing program for maintenance on vehicles; designed to provide routine care, oil changes and the like, as well as catch minor problems before they become major ones

Many of the ongoing operational safety objectives can be accomplished by having a strong **preventive maintenance (PM) program** (see Figure 4–6). The PM program should include daily checks of emergency lights, fluids, and the operation of vehicle components including brakes, pumps, transmission, and aerial ladders. The daily check should be performed by the apparatus operator assigned that day. The operator should be trained to look for maintenance problems before they become worse or cause an accident. Using the apparatus operator also allows the operator to become familiar with the vehicle and vehicle components.

A more thorough check should be performed weekly, including running small tools, cleaning the apparatus thoroughly, checking tire wear and pressures, and back-flushing pumps. A monthly check may go even further, requiring the performance of such tasks as changing water in the water tanks.

Another component of the PM program is a semiannual trip to the maintenance professional. This trip can provide for scheduled fluid changes and visual checks of brake linings and other critical vehicle systems. In the case of fire apparatus, pumps and ladders must be tested annually as well.

The *NFPA* standard *1915* requires the organization to develop and implement a schedule of service and maintenance for fire apparatus based on the manufacturer's recommendations, local conditions, and operating conditions. It also describes the requirements for various systems and components of the fire apparatus. *NFPA 1915* further requires that the organization develop written criteria for when the apparatus is to be taken out of service. It is required that this criteria consider 49 *CFR*, part 390, "Federal Motor Carrier Safety Regulations"; any applicable federal, state, and local regulations; applicable nationally recognized standards; manufacturers' recommendations; and guidelines established by the organization or its designated service and maintenance staff. The apparatus cannot be returned to service until after defects and deficiencies have been corrected. An example policy is shown in Text Box 4–3.

Text Box 4–3   *Vehicle out of service guide.* (Courtesy Palm Harbor Fire Rescue.)

PURPOSE: To keep our vehicles in safe operating condition and to assist personnel in placing vehicles out of service. This SOP categorizes apparatus repair needs utilizing a red, yellow, and green format of all items that require repair based on seriousness of mechanical or physical condition.

RED (IMMEDIATE): Any and all items that would cause a vehicle to be removed from service immediately. Prompt notification of the on-duty officer and shift commander.

A. SAFETY: Brakes, broken belts, elimination of a major component such as an alternator, power steering, water pump, transmission, flat or heavily damaged tire, audible air leak at brake chamber, loss of emergency lighting or more than one signaling device, etc.

B. INCAPACITATING: Any problem found that prevents starting or does not allow vehicle to operate properly after start up.

C. EXPANDING: Any problem, large or small, left unattended that might potentially cause serious damage to the vehicle or any of its major operating systems; items such as serious leaks (i.e. excessive engine oil, transmissions, hydraulic oil, transfer case, rear axle, wheel bearings, antifreeze, etc.). Alternators not charging to full capacity.

D. ACCIDENT: Any accident in which the vehicle has sustained heavy damage.

YELLOW (AS SOON AS POSSIBLE BUT NOT IMMEDIATELY): Any and all items that would not cause a vehicle to be removed from service. Requires notification of the on-duty officer and shift commander.

A. Any item that is important and needs prompt attention, but does not require the vehicle to be taken out of service immediately. (Non audible slow air leak, worn belts, etc.).

B. Any problem, large or small, left unattended or that may cause serious damage to any of the vehicle's components if not repaired before next scheduled service.

C. Warning devices (door ajar alarms) not working properly.

GREEN (CAN WAIT, OR IN-HOUSE REPAIR): Any item that should be repaired that cannot be repaired by the driver engineer or shift personnel, but can wait for the next scheduled maintenance appointment. These items need to be documented on the back sheet of the check-off sheets.

A. This would be such items as dash lights, torn upholstery, worn tires, marker light covers, leaking pump packing, or any nonessential components on vehicle that will not hinder operation of the vehicle or cause damage to it by not being repaired immediately.

National Association of Emergency Vehicle Technicians (NAEVT) an organization that bestows professional certification in many areas for persons involved in emergency service vehicle maintenance

This type of preventive maintenance program can be applied to any size or any type of department. Whether the apparatus is a fire truck or an ambulance, the same concepts apply. Lack of preventive maintenance reduces the useful life of a vehicle, increases maintenance costs, and can have a significant negative effect on the safety of those relying on the vehicle.

Standard operating procedures and training can also help in the day-to-day operational safety process. Procedures identifying the minimum number of personnel to perform certain tasks have been successful in reducing injuries. For example, for reloading the hose, the procedure may require four persons to be on hand: two in the hose bed, one on the back step, and one on the ground. Keeping the apparatus running board free from debris or fluids prevents slip and fall injuries. Proper lighting when working around the vehicle can also reduce injuries.

Apparatus safety procedures should also address hearing conservation. The USFA has published a manual to assist emergency service organizations in developing a hearing conservation program.

Apparatus safety must include risk identification and evaluation, and the implementation of control methods that can only be designed at the local level. Text Box 4–4 outlines some apparatus safety considerations, although some are related to response. Another resource is the **National Association of Emergency Vehicle Technicians** (NAEVT), which can provide valuable information on apparatus maintenance and safety operating practices. The address is listed in the further resource section of Appendix B.

Text Box 4–4 *Example policy.* (Courtesy Palm Harbor Fire Rescue)

APPARATUS SAFETY

355.0 Purpose

The purpose of these standards is to establish safety procedure in regard to apparatus practices thereby reducing the potential of accidental injury or death to fire department personnel and the general public.

355.1 Vehicle and Equipment Inspection Procedures

A. In order to ensure vehicle readiness and presence of essential equipment, all apparatus and equipment shall be regularly inspected and/or tested by personnel at intervals prescribed on the existing Vehicle Inspection Form. Daily vehicle and equipment checks shall be performed as close to 0800 hours as practicable and their timely completion is to be superseded only by emergency calls or by the order of the officer in charge.

B. Any apparatus or equipment deficiencies shall be noted on the departmental Vehicle/Equipment Deficiency Form. The officer in charge shall also be verbally notified of any such deficiencies. The officer in charge shall then take necessary action to effect repairs or remedy the deficiency. The equipment or apparatus shall be placed out of service should the deficiency be deemed serious or hazardous.

C. All vehicles and equipment shall be outside of the bays while operating (weather permitting).

355.2 Driver's Responsibility

A. It is the responsibility of the driver of each fire department vehicle to drive safely and prudently at all times. Vehicles shall be operated in compliance with traffic laws pursuant to F.S. 316. Emergency response does not absolve the driver of any responsibility to drive with due caution. The driver of the emergency vehicle is responsible for its safe operation at all times. The officer in charge of the vehicle has responsibility for the safety of all operations.

B. It is the driver's responsibility to ensure that all passengers are safely aboard the apparatus and seated prior to moving the vehicle.

C. When responding to an emergency, emergency warning lights must be on and sirens must be sounded to warn drivers of other vehicles.

D. Intersections present the greatest potential danger to emergency vehicles. When approaching and crossing an intersection with the right-of-way, drivers shall not exceed the posted speed limit.

  When approaching a negative right-of-way (red light, stop sign, or yield), the vehicle shall come to a complete stop. The vehicle may proceed only when the driver can account for all oncoming traffic and all lanes yielding right-of-way.

E. In order to avoid unnecessary emergency response, the following rules shall apply:

  1. When the first unit reports on the scene with "nothing showing," "investigating," or the equivalent, any additional units shall continue emergency, but shall not exceed the posted speed limit and shall come to a complete stop before proceeding at every negative right-of-way intersection.

  2. The first arriving unit will advise additional units to respond nonemergency whenever appropriate.

F. During an emergency response, fire vehicles should avoid passing one another. If passing is necessary, arrangements should be conducted through radio communications.

G. The unique hazards of driving on or adjacent to the fireground require the driver to use extreme caution and to be alert and prepared to react to the unexpected.

H. When stopped at the scene of an incident, vehicles should be placed to protect personnel who may be working in the street and warning lights shall be used to make approaching traffic aware of the incident. At night, vehicle-mounted floodlights and any other lighting available shall be used to illuminate the scene.

I. Backing. When the backing of apparatus is unavoidable, the driver shall use a guide to back and shall not proceed to back unless the guide is clearly visible in the mirror(s) being used for visualization. Illumination of the backup area with floodlights and/or flashlights shall be used in times of darkness or limited visibility. If no guide is available, the driver shall dismount and walk completely around the apparatus before proceeding to back. Audible backup alarms shall be used during any backing procedure.

355.3 Seating/Seat Belts

A. Personnel are to ride in a seated position at all times that the apparatus is in motion and only in seats or positions where seat belts are provided. Seat belts are to be utilized at all times when vehicles are in motion. Under no circumstances will personnel be permitted to ride in a standing position or on any sideboards or tailboards of any apparatus.

B. EXCEPTIONS to 355.3(A) above:

1. Seat belt use will not be required during actual fire combat from exterior positions on the brush or water trucks during brush fires. For this reason, brush/water truck drivers shall exercise extreme caution with regard to speed, terrain, overhead clearance, etc. If terrain is extremely rough or other conditions exist that would render exterior positioning unsafe, the rider shall dismount from the vehicle and either proceed on foot or take a seated position in the interior of the cab. This exception does not exempt personnel from the required use of seat belts in brush/water apparatus during routine or emergency over-the-road operations or while traveling point-to-point during off-road operations.

2. Fire department personnel riding in ground or air ambulances or any other vehicles shall wear seat belts when possible.

3. It is obvious that seat belts are impractical during repacking of hose in the hose bed of apparatus. This should not be attempted without adequate personnel.

C. When responding to calls requiring the donning of protective clothing, donning shall be completed prior to response since this procedure would be difficult to complete while seat belts are engaged.

D. If a call requiring the donning of protective clothing is received during routine driving of apparatus, the driver shall proceed to a safe location off the roadway where personnel may don protective clothing prior to continuing response. It is not recommended that personnel wait until arrival at the scene to don protective clothing.

E. Donning of air packs while apparatus is in motion shall be done only in apparatus provided with seating which permits donning while seat belts are engaged and proper seating/posture can be maintained. In apparatus where this is not possible, personnel shall don air packs upon arrival at the scene.

F. Safety bars (e.g., Man Savers) shall be in the down position at all times and shall not be disabled or altered in such a manner that would defeat intended function.

G. A common cause of ankle/knee injuries is forward (front-facing) dismount from apparatus. All personnel, especially when wearing air packs or carrying heavy loads, should back down off of apparatus using hand-hold handles to assist.

■ **Note**

Apparatus driver selection should consider the human aspects such as attitude, knowledge, mental fitness, judgment, physical fitness, and past driving record.

RESPONSE SAFETY

The first consideration in response safety is the selection, training, and capabilities of the driver operator. Apparatus driver selection should consider the human aspects such as attitude, knowledge, mental fitness, judgment, physical fitness, and past driving record. *NFPA 1002 Standard for Fire Apparatus Driver/Operator Professional Qualifications* outlines the requirements for operators of fire apparatus in both emergency and nonemergency situations. The standard requires that the driver be licensed to operate the vehicles they are required to drive and requires a medical exam to prove fitness to operate an emergency vehicle. The specific requirements and performance objectives for driving and operating various types of apparatus can be found in the standard. The various types of apparatus include those equipped with an attack or fire pump in Chapter 5, aerial devices in Chapter 6, tiller operators in Chapter 7, wildland fire apparatus in Chapter 8, aircraft rescue and firefighting apparatus in Chapter 9, and water supply apparatus in Chapter 10. The standard also requires the organization to evaluate these required performance objectives in the selection of drivers.

In order to meet *NFPA 1002*, most departments perform some type of testing prior to driver selection, which measures some or all of these characteristics. Further, prior to or after selection, the driver should be trained in emergency vehicle operations in a vehicle similar to the one he or she will have to drive and operate. This training program should address legal issues, physical forces including controlling the vehicle on a variety of road surfaces, safe stopping distances, vehicle maintenance, departmental safety procedures, general safe driving practices, and specific safe driving practices based on the geographic area and environment. This training must have both classroom and practical components. A driving course is a common approach to the practical phase of this training, but provisions must be made for different driving surfaces, including wet roads. A commonly used driving course setup is provided in Figure 4–7.

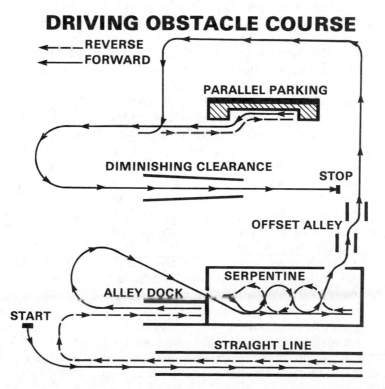

Figure 4–7 *A typical obstacle course for emergency services driver training.* (Courtesy USFA.)

The department's insurance company may also offer driver training resources; some even have complete programs including an instructor. The emergency vehicle driver must have the capabilities to control the vehicle and react quickly and appropriately to changing situations. Excessive speed, reckless driving, and failing to slow down or stop going through intersections have been the causes of many emergency vehicle accidents. Policies that prohibit operating an emergency vehicle while under the influence of drugs or alcohol, including prescription medications, must also be in place and enforced.

Once the driver is selected and trained, the program must focus on the response. Consideration for procedures that require the use of seats and seat belts is necessary. The habit of riding tailboards or sidesteps is unacceptable and for the most part has stopped. However, standing in the jump seat area still occurs. The supervisor on the emergency vehicle should be required to ensure that all occupants are properly seated and belted prior to the response.

The type of response also has safety implications. An emergency response using warning devices is more dangerous than a response in which the vehicle flows normally with traffic. The safety manager, in conjunction with the department management, should review the types of incidents that the department responds to and determine which may not require emergency response. The St. Louis fire department (SLFD) has taken this approach and reduced the number of intersection accidents greatly. Since implementation of this program, the SLFD has only had one serious accident in three years or 180,000 alarms. Text Box 4–5 describes this program.

The environment in which the response takes place can also be analyzed and risks identified. Traffic patterns and common response routes should be studied. Apparatus accidents have occurred between two fire department vehicles responding to the same incident from different locations. This risk can be reduced with better radio communication and preincident procedures for response. Target intersections can also be identified. A problem intersection where emergency vehicle collisions or near misses have occurred should be avoided. However, this choice is seldom reasonable. The department could look at intersection control. Traffic signal preemption is popular across the county. There are several types of systems available. In general the traffic signal controller gets a signal from the preemption device and changes the traffic signal light to green for the direction that the emergency vehicle is traveling and red in all other directions. The signal from the preemption device can come from an optical emitter on the response vehicle, the vehicle's siren, and if the intersection is near the station, it can be hardwired to the station. The use of global positioning system (GPS) for traffic preemption should also be available soon.

Procedures may require that the vehicle come to a complete stop at every intersection or not exceed the speed limit in residential neighborhoods. The environment is dynamic, traffic patterns change with time of day, weather can change quickly. Road construction should be planned for and response routes evaluated. Areas along the response route that have poor visibility should be identified prior

Text Box 4–5   *St. Louis Fire Department nonemergency response policy.*

Companies and chief officers will respond on the quiet, no lights or sirens, to the following incidents:

- Automatic alarms
- Sprinkler alarms
- Natural gas leaks
- Wires down
- Calls for manpower
- Flush jobs
- Lockouts
- Investigation drums, barrels, unknown odors, and such
- Carbon monoxide detector alarms
- Rubbish, weeds, and dumpster fires
- Move-ups to city or county fire stations
- Broken sprinkler or water pipes
- Smoke detectors
- Manual pull stations
- Refrigerator doors
- Plugging details
- Assist the police
- Keys in running vehicles

If a response is dispatched on the quiet and additional information is received by fire alarm indicating that life is in danger, persons are injured, there is a working fire, and so forth, fire alarms will upgrade the response to "urgent," lights and sirens.

---

to the response. Railroad crossings, particularly those that are not controlled should be noted on maps, or the drivers should be trained in their locations. In 2002, a volunteer firefighter/engineer was fatally injured when the tanker truck he was driving was struck by a freight train as he attempted to traverse a private, unguarded railroad crossing. The victim was returning to the station from a live-burn training exercise. NIOSH made the following recommendations in its report of the incident:

- Ensure that drivers of fire department vehicles come to a complete stop at all unguarded railroad grade crossings during emergency response or non-emergency travel
- Revise or develop SOGs for safely driving emergency vehicles during emergency response and nonemergency travel addressing hazards firefighters are likely to encounter, such as railroad crossings

- Enforce policies that require that all firefighters who ride in emergency fire apparatus are belted securely by seat belts
- Provide firefighter training on railway traffic safety in communities such as this in which a high density of railway traffic exists

PREINCIDENT PLANNING

The value of preincident planning cannot be underscored enough, not only from a safety standpoint, but also from an operational standpoint. Having preincident plans available is like a coach having the playbook at a game. In the case of a fire, preincident planning allows the fire department to fight the fire before it happens. From a safety standpoint, preincident planning can be used to identify and evaluate potential risk long before an incident occurs. For example, preincident planning may reveal that an occupancy uses and stores combustible gases. The gases are compressed and stored in a storage room. Having identified this hazard, responders can adjust tactics to limit their exposure to potential explosions or **boiling liquid expanding vapor explosions** (BLEVES), should a fire occur.

NFPA 1620, Recommended Practice for Pre-Incident Planning outlines the necessary components for a preincident planning program to be effective. It covers various types of occupancies as well as the varied kinds of information that should be collected. Before implementation of a comprehensive preincident planning program, this document should be reviewed.

To be effective, the preincident planning process should contain the following components:

- The preincident plan should be on a form used departmentwide.
- Preincident planning should be done by the responders so that they can become familiar with the building during the preparation of the plan.
- The process should provide for updating the plan at given time intervals. A good goal would be an annual update. Remember, occupancies change and renovations occur.
- Target hazards should be identified and receive priority in the planning process. Target hazards may include:
  - Health care facilities
  - Large industrial occupancies
  - Facilities using or storing hazardous materials
  - High-rise buildings
  - Malls or other high occupancy locations, such as office complexes or stadiums
  - Schools
  - Hotels, motels, and apartment buildings

- The preplan should include text as a reference, a site plan, and a floor plan.
- The preplan should show or provide the following information:
  - Location
  - Construction features
  - Building emergency contacts
  - Occupancy type and load
  - Hazardous materials or processes
  - Location of utilities
  - Location of entrances and exits
  - Ventilation locations
  - Nearest water supply
  - Need for specialized extinguishing agents
  - Private fire protection equipment and systems on site, including fire department connections
  - Mutual aid plans
  - In addition, some preplans provide for placement of apparatus on a first-alarm assignment.

Although preincident planning components address many of the needs of responders to fires or hazardous material emergencies, preplanning also applies to EMS. For example, EMS responders may want to preplan alternate accesses to a nursing home after hours. Are there alternate means of egress? Will a stretcher fit in the building's elevator? What about access to locked apartments for a medical emergency when a person cannot make it to the door. Many times complexes have on-site maintenance or security with keys, which is easier and safer than breaking down the door and does not upset the owner as much.

Preincident planning is an excellent tool when considering preincident safety and risk identification. However, preincident planning is commonly limited to target hazards and many of the profession's injuries and deaths occur in single-family dwellings.

SAFETY IN TRAINING

Based on the 2002 NFPA injury report, 7,600 firefighter injuries occurred during training. This number represents 9.4% of the total injuries. In terms of firefighter line-of-duty deaths, according to the 2002 USFA report, eleven firefighter deaths occurred during training activities. This was 11% of the total deaths of 100. Training evolutions are generally created to simulate actual events, but the training evolutions must be controlled so that injuries can be further prevented. *NFPA 1403, Standard on Live Fire Training Evolutions*, was created as a result of a number of injuries and deaths that occurred during live-fire training. The standard followed closely the deaths of three Milford, Michigan, firefighters who were killed during a training burn at an acquired structure. The firefighters were engaged in a training exercise in a two-story structure and were trapped on the second floor after a fire was started on the first floor. They did not have adequate personal protective equipment (PPE), nor the safety of a handline. Training evolutions must be designed and conducted only after a thorough risk analysis including identification and control of the risks. Injuries and deaths associated with training are avoidable, and the safety manager must adopt a zero tolerance level in this regard.

NFPA 1403 sets forth procedures and requirements for live-fire training, some of which are also applicable to nonfire training evolutions. The *1403* standard addresses the following subjects:

- Assignment of safety officers
- Preburn walk-through of the structure
- Building acquisition procedures
- Securing utilities
- Environmental impact of the burn
- Building construction and conditions
- Use of proper protective equipment
- Proper water supply including backup lines and supply
- Use of an incident management system
- Instructor-to-student ratio
- On-scene emergency medical care

NFPA 1410, Standard on Training for Initial Emergency Scene Operations describes the minimum requirements for evaluating training for initial fire suppression and rescue operations of a fire department and the personnel engaged in emergency scene operations. It has been designed to provide fire departments with an objective method of measuring performance for initial fire suppression and rescue procedures using available personnel and equipment. Although not directly related to safety, using *NFPA 1410* can support safe practices by ensuring that evolutions can be completed safely using adequate numbers of personnel and proper equipment.

Many of the requirements addressed in these standards can become part of a department's standard procedures for training evolutions. The use of protective clothing for example—how often is it permitted that hose streams are operated with just helmets or helmets and gloves, or worse yet, nothing? In this case, the training is not simulating actual conditions. Furthermore, the risk of injury from the hose is much greater than if the member was in full PPE. Is it department policy to assign a safety officer during a hazardous materials exercise, or during a multicompany drill at the drill tower using ladders. It should be policy. Do your policies provide for rehabilitation and EMS at the site of outside evolutions?

Do you follow the same safety procedures during EMS training as you would in the field in terms of disposal of blood products and needles? How about protective equipment during extrication exercises? Repeating cutting up cars in the summer in full PPE tends to make one cut corners. Having a safety officer watching and prohibiting such actions may save an injury. The procedures must also provide for an accountability system at multicompany training evolutions similar to that used at real incidents. This not only allows for practicing the system but also affords another level of safety for the participants.

■ Note
The procedures must also provide for an accountability system at multicompany training evolutions similar to that used at real incidents.

EMPLOYEE WELLNESS AND FITNESS

An employee wellness and fitness program is an essential component of the safety and health plan. Fitness and wellness programs can be developed locally using in-house talent or by contracting with professional exercise and wellness consultants. A number of resources are available to assist departments in developing these programs. The IAFF/IAFC has a joint initiative, the USFA publishes a fitness coordinator's manual, local universities often have an industry hygienist who may be able to help, and department physicians may be able to provide information on program development. A number of components are necessary for a program to be successful and effective.

It is difficult to separate the terms *wellness* and *fitness*. Generally the occupational safety and health programs each have components that can fit in these categories, but for the program to be effective there must be tight integration. Wellness is a broad term that would include components relating to medical fitness, physical fitness, and emotional fitness. Wellness programs should also include behavior modification programs including weight reduction, tobacco use cessation, nutrition, and stress reduction.

A successful wellness and fitness program must also have provisions for after-injury rehabilitation and care. A rehabilitation program should be designed to ensure that the employee has a path to rehabilitation and return to duty with the employee's safety in mind. Rehabilitation programs should include input from the department doctor and physical therapy agencies familiar with the job of an emergency responder. **Alternative duty programs** are also helpful to permit a safe return. These programs allow an employee to work during his or her recovery, but within restrictions based on physician's orders. Periodic evaluations of the employee's condition are also an integral part of the program.

alternative duty programs
sometimes called light duty or modified duty, these programs allow an injured employee to return to work, with restrictions, for some period of time while recuperating

Medical Fitness

Medical fitness can be assessed by performing yearly medical exams. While recommendations differ as to how often these exams should be conducted for different age groups, it is agreed that annual medical exams are a necessity. *NFPA 1582*

aerobic fitness
a measurement of the body's ability to perform and utilize oxygen

cardiovascular fitness
fitness levels associated with the cardiovascular system, including the heart and circulatory system

muscular strength
the maximal amount of force a muscle or group of muscles can exert in a single contraction; the ability to apply force

muscular endurance
the ability of the muscle to perform repeated contraction for a prolonged period of time; the ability of the muscle to persist

flexibility
the range of motion in a joint, which depends on the extensibility of soft tissues

body composition
a measure to show the percentage of fat in the body; there are certain published parameters for what is considered average or normal

addresses the requirements for a comprehensive occupational medical program for fire departments. This standard requires a medical evaluation at the time of employment (prior to assignment) and annually thereafter. The medical exam should be conducted by a physician who is very knowledgeable in occupational issues and specifically with the essential job tasks, which are outlined in the standard. The standard provides valuable guidance for the fire department physician. The standard also identifies the essential tests and components of the medical exam. Text Box 4–6 provides the tests recommended by the IAFF/IAFC wellness-fitness initiative and are consistent with the requirements of *NFPA 1582*. *NFPA 1582* lists conditions for each body system and categorizes problems into either category A or B. Category A conditions are defined in the standard as those conditions that *would* preclude a person from performing tasks in a training or emergency operational environment because it would present a significant risk to the safety and health of the person or others. The fire department physician shall not certify for duty those persons with category A conditions. Category B conditions are defined in the standard as those that *could* preclude a person from performing tasks in a training or emergency operational environment by presenting a significant risk to the safety and health of the person or others, but does not automatically preclude a person from becoming certified as meeting medical standards for duty.

Physical Fitness

According to *NFPA 1583, Standard on Health-Related Fitness Programs for Fire Fighters*, annual physical fitness assessments for emergency responders should address five components: **Aerobic** or **Cardiovascular fitness**, **muscular strength**, **muscular endurance**, **flexibility**, and **body composition**. These components interrelate and are required of the emergency responder. A person can have a great deal of muscular fitness and be very strong, but have a low level of cardiovascular fitness or endurance. Clearly the work of an emergency responder requires both.

Cardiovascular fitness and body composition can be assessed as part of the medical exam. Muscular fitness can be assessed using a battery of tests. The hand-grip dynamometer can be used to measure grip strength, the arm-flex dynamometer measures the maximum force generated from arm flexion, and the leg-extension dynamometer measures the maximum force that can be generated from extension of the legs. Push-ups and sit-ups can also be used to measure muscular strength and endurance.

Once the fitness level is identified, a program should be developed by the department fitness coordinator on an individual basis. For example, some employees may lack cardiovascular endurance, while others may have excellent cardiovascular endurance but lack muscular fitness. The program should also provide for on-duty time and the facilities for members to work out and maintain the required level of fitness.

Text Box 4–6   *Medical tests recommended by the IAFF/IAFC wellness-fitness initiative.*

Medical History Questionnaire

Vital Signs

Hands-On Physical

  Head, eye, ears, nose, and throat

  Neck exam

  Cardiovascular—submaximal tests

  Pulmonary—including spirometry and chest X ray (repeat X rays recommended every 3 years)

  Gastrointestinal

  Genitourinary

  Rectal—including digital rectal exam

  Lymph nodes

  Musculoskeletal

  Body composition

Laboratory Tests

White blood cell count	Potassium
Platelet count	Albumin
Glucose	Heavy metal screening
Sodium	Red blood cell count
Total protein	Triglycerides
Cholesterol tests	Creatine
Differential	Carbon dioxide
Liver function tests	Calcium
Blood urea nitrogen	Urinalysis

Vision Test

Hearing Test

Cancer Screening

Breast exam	Fecal occult blood testing
Prostate specific antigen	Pap smear
Testicular exam	Skin exam
Mammogram	

Immunizations and Infectious Disease Screening

Vaccinations	Varicella
Hepatitis C	Hepatitis B virus
HIV	Polio
Tuberculosis	

Furthermore, the program must consider the following five basic components necessary for an exercise plan:

1. The mode or type of exercise
2. The intensity or difficulty of the exercise
3. The duration or length of the exercise session
4. The frequency or number of sessions per day or week
5. The progression, meaning the gradual increases in workload to promote a training adaptation

Emotional Fitness

Emotional fitness is also a necessary component of the program. A responder who is under undue stress, from whatever cause, will not be able to perform effectively and is prone to injury. Emotional fitness can be improved by having programs in place for employee assistance. These programs should be accessible to employees and their families on a confidential basis. Often employers pay the cost for a fixed number of visits through a contract with local behavior care providers. Family members are encouraged to participate too, as often the emotional problems are related to family and home life. Services provided by some employee assistance providers include:

> Drug or alcohol abuse treatment
>
> Tobacco use cessation
>
> Family problems
>
> Financial problems
>
> Stress management
>
> Critical incident stress management

■ Note
Employee assistance programs may also be part of the discipline process in an organization.

critical incident stress management (CISM)
a process for managing the short- and long-term effects of critical incident stress reactions

Employee assistance programs may also be part of the discipline process in an organization. If an employee's work performance takes a sudden downturn, the reason may be an emotional problem. Requiring mandatory participation in an employee assistance program can be used as a step in the discipline process and contribute to the future of the employee and the department.

Clearly a factor in the emotional wellness of an emergency responder is **critical incident stress management** (CISM). Having a CISM program in place contributes to the health and safety program. This subject is presented in Chapter 8 with postincident safety issues.

INTERAGENCY CONSIDERATIONS

Another preincident safety and health issue that is sometimes overlooked is the relationship between the responder's agency and other agencies with which the organization cooperates. While these relationships have always had operational and safety implications, the tragic events in New York and Virginia on September 11, 2001, as well as the events that followed involving anthrax brought the need for interagency coordination to national attention. After these events, President Bush issued Homeland Security Presidential Directive 5 which purpose was to enhance the ability of the United States to manage domestic incidents by establishing a single, comprehensive national incident management system.

In the forefront of interagency relationships is that with mutual or automatic aid providers. In most emergency response agencies, some form of mutual/automatic aid agreements exists. The safety program manager must know how the other departments operate. Are they safety aware, do they use the same incident management system, the same accountability system? Do their operations comply with recognized standards? *NFPA 1710* requires certain operational issues to be included in mutual/automatic aid agreements. Are the communications systems compatible? Communication system compatibility was also an issue in the 2001 terrorist attacks.

Responders also frequently work with law enforcement agencies. In most cases, the emergency responders referred to in this text are not law enforcement officers. That creates the need for integration in response to some incidents and a preincident understanding of roles and relationships. Law enforcement agencies often respond to many incidents with the fire or EMS responder including vehicle crashes, incidents requiring crowd or traffic control, incidents resulting from criminal activities, and in some jurisdictions as first responders to all medical incidents. In some areas, the law enforcement agencies may provide rescue or aeromedical evacuation. This response creates the need for integration in response to some incidents and a preincident understanding of roles and relationships (see Figure 4–8).

Having an understanding of each agency's priorities and roles prior to an incident occurring is critical to safety and health activities and must be preplanned. In some situations there will not be clear lines of authority and responsibility. In these situations, the incident management system must provide for **unified command**, which provides for an incident commander from both law enforcement and the emergency response agencies. These commanders work in unity to bring the incident to a safe conclusion. Other issues can be preplanned; for example, do the emergency responders wait for police clearance before entering a scene where violence such as shootings or stabbings is possible? Do law enforcement agencies provide for traffic control at an interstate accident, or is their priority to reopen the highway? What are the expectations of the emergency responder agencies in terms of protection from law enforcement during periods of civil unrest?

This relationship is a two-way street. Law enforcement may ask the emergency responders to avoid using reds lights and sirens when nearing a scene of a shooting in the street in order not to draw a crowd, making the officers' job unsafe. Or the aeromedical helicopter operated by the police may need some specific ground support for landing zones. All of these considerations can be dealt with on

unified command
used in the IMS when two or more jurisdictions or agencies share incident command responsibilities but do so in conjunction with each other

Figure 4–8 *At this vehicle crash, three different public safety agencies are working together.* (Photo courtesy of Firefighter/ Paramedic Chris Eisenhardt.)

■ **Note**

The important point to realize is that the issues must be decided on prior to the incident and with the safety of *all* responders in the forefront of concern.

a local level and in many different ways. The important point to realize is that the issues must be decided on prior to the incident and with the safety of *all* responders in the forefront of concern.

In terms of interagency cooperation, emergency responders will probably interact most with mutual aid and law enforcement, but other agencies are involved as well. Hospitals provide initial treatment or medical surveillance after exposures to infectious diseases. The relationship with the local emergency room should be an integral part of safety and health programs. Again, what are the expectations of the emergency responder agencies and the expectations of the hospital? Depending on the department situation, a communication center may be operated by a separate agency wherein safety and health concerns for responders can be preplanned. The local health department can be a resource for information on infectious diseases. The health department's role has become more important to emergency responders as the threat of chemical, biological, radiological, nuclear, or explosive attacks on the country have increased. The health department has many resources available for these events, and in fact may be the lead agency depending on the incident. During a CBRNE incident, law enforcement, health, and the fire department may use a unified command as shown in Figure 4–9.

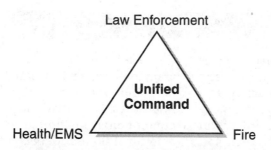

Figure 4-9 *Interagency incident management considerations—the command triangle—fire, law enforcement, and health/EMS—for expanded unified command. (From the Florida Field Operations Guide (FOG))*

Depending on local conditions, safety and health for the emergency responder can be impacted, either positively or negatively, by a preincident relationship with agencies other than the responder's. This relationship can go far not only from a safety and health standpoint, but also from the standpoint of effective and integrated operations.

SUMMARY

Preincident safety encompasses a number of components including station safety, apparatus safety, response safety, preincident planning, safety during training, wellness/fitness programs, and interagency relations. Station and apparatus safety can be subdivided into issues relating to design and ongoing, day-to-day operations. Response safety includes driver selection, driver training programs, and policies affecting response. Preincident planning is a tool that can be used by responders to effect safer operations. Building characteristics, occupancy, and particularly hazards can be identified prior to an incident. Means of access and tactical considerations can also be discussed prior to an emergency and under better conditions.

Safety during training is a concern in the overall health and safety program. Training is often a simulation of real conditions and is a controlled, although somewhat unpredictable, environment. All training evolutions should be conducted with participant safety as the number one priority.

An integral part of the safety and health program is a program to ensure the highest level of fitness and wellness for the responders. Wellness is a broad term that encompasses a number of components, including medical, physical, and emotional fitness. Medical fitness can be assessed with medical exams. Physical fitness assessments are necessary and individual programs are developed for responder improvement. An employee assistance program is vital in ensuring emotional fitness.

Emergency responders work alongside and rely on a number of other agencies. Strong interagency coordination and preincident defined roles and expectation can be helpful in preventing injuries and death. Interagency coordination also leads to a more effective operation.

Concluding Thought: Safety is a state of mind. Many components can be prepared for before an incident.

REVIEW QUESTIONS

1. Which of the following would not normally be considered as part of the daily vehicle apparatus checks?

   A. Back flushing the pump

   B. Checking the emergency lights

   C. Checking the oil level

   D. Checking the seat belts

2. List the areas of concern when selecting emergency vehicle driver/operators.

3. Preventive maintenance programs should require biannual trips to the maintenance professional.

   A. True

   B. False

4. What is one option that exists for traffic light control at intersections?

5. List five of the lab tests that should be included in an annual medical examination.

6. Emotional fitness is one of three types of fitness important to the emergency responder.

   A. True

   B. False

7. List two external agencies with which a department may want to coordinate before an incident.

8. For the purpose of preincident planning, which of the following would not be considered a target hazard?

A. A hospital

B. A lumber yard

C. A railyard

D. A two-unit duplex

9. List three components of an effective vehicle preventive maintenance program.

10. List three methods of determining muscular fitness.

ACTIVITIES

1. Review your department's policy on medical examinations. Compare the policy to the standard and the recommendations in this text. How does your exam compare? If you do not have a medical examination program, research programs and write a draft policy.

2. Review your department's standard operating procedures for response to incidents. Is the policy current? Could the emergency response mode be reduced for some incidents? Is there a policy or procedure for the selection and training of drivers?

3. Review your department's standard procedures for safety in training. Does your live-fire training comply with *NFPA 1403*? With application from *1403*, could improvements be made to day-to-day training activities?

4. Identify your department's relationship with other agencies. Are there standard procedures or other prearranged agreements? Are the policies in effect or do they need to be established?

Chapter

5

SAFETY AT THE FIRE EMERGENCY

Learning Objectives

Upon completion of this chapter, you should be able to:

- List the three incident priorities.
- Explain the relationship between the three incident priorities and the relationship to health and safety.
- Discuss in general terms the hazards faced by responders to fire incidents.
- List the components of personal protective equipment used for fire incidents.
- Discuss the need for and the components of an effective accountability system.
- Discuss the types of and the relationship between incident management systems and health and safety of the responder.
- Define the need for and uses of a rapid intervention team.
- Discuss the components of incident rehabilitation.

CASE REVIEW

On Valentine's Day, 1995, at 12:22 A.M., the Pittsburgh Fire Department responded to a house fire in the 8300 block of Bricelyn Street. On arrival, the firefighters found a fire in a three-story wood frame structure with a basement. The building was constructed on a grade, which gave the appearance from the front that the house was only two stories. The initial tactic was to mount an interior attack and the first arriving engine stretched a line to the interior to search for the fire and to effect extinguishment.

During this operation, a stairwell used by the firefighters collapsed. After the collapse, firefighters were reported to be trapped inside. Because several companies had by then arrived and were working in the interior, supervisors were unable to quickly assess which companies were missing.

Shortly after the report of firefighters being trapped, several firefighters were rescued through an exterior window in the rear. At this time there was no report accounting for all firefighters on the scene, and, as a result, the incident commander was not aware that three other firefighters remained trapped in the building. The three remaining firefighters were discovered after most of the fire was knocked down and smoke ventilated, about 1 hour after the firefighters entered the building.

After the NFPA investigation was performed, several factors were identified as contributing to this tragedy. Two of the factors cited were lack of procedures that allow for a quick accounting of firefighters from the beginning of the incident and the use of rapid intervention teams for unexpected situations that occur on the fire scene, enhancing the chance of rescue and survival.

INTRODUCTION

As seen in Chapter 1, most firefighter deaths and injuries occur while operating at fire scenes as a result of different causes, ranging from stress to building collapse. The safety and health program manager must have a good knowledge of the hazards faced at fires and procedures that will ensure safer operations.

Forming an understanding of the incident priorities and their relationship with operations and hazards will improve the safety manager's ability to design a training program, purchase equipment, and implement procedures targeted toward injury reduction.

INCIDENT PRIORITIES AND SAFETY

Priorities for incident management can be categorized and placed in descending order of seriousness into life safety, incident stabilization, and property conservation. These priorities are the first thing that the incident commander must consider in the incident, and they should be continuously reevaluated as the incident progresses.

Life Safety

Life safety is the first incident priority as it is based on the premise that life safety must be maximized. Life safety is the group of activities that ensures that the threat of injury or death to civilians and emergency personnel is reduced to the absolute minimum. This is done by limiting the exposure of danger to the absolute minimum. In a simple situation, it may mean that emergency responders wear protective clothing while operating. In a complex incident involving a fire in several buildings, it may require the evacuation of several blocks of residences and setting up a collapse zone. Regardless of the situation or how the priority is met, life safety is always number one throughout the incident. Life safety activities must include and consider responder safety.

Incident Stabilization

The second priority is incident stabilization. During this time, attempts are made to solve the problem. Incident stabilization is the group of activities required to stop additional damage or danger. Stabilization could mean a quick interior attack on a room and contents fire or a defense attack on a fire to stop the spread from building to building. Size-up helps to determine initial strategy (see Figure 5–1 and Figure 5–2).

Property Conservation

Property conservation is the final incident priority and involves reducing the loss to property and the long-term health and welfare of the people affected. This group of activities is commonly called salvage or stopping the loss (see Figure 5–3).

Figure 5–1 *Safety officer should perform a size-up based on incident priorities.*

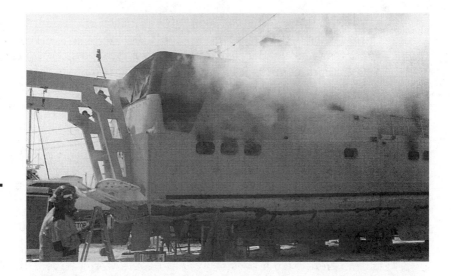

Figure 5–2 *Fire officer performs size-up on arrival at the incident scene.*

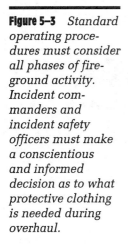

Figure 5–3 *Standard operating procedures must consider all phases of fireground activity. Incident commanders and incident safety officers must make a conscientious and informed decision as to what protective clothing is needed during overhaul.*

In a simple incident, the use of a salvage cover may be all that is necessary. In a large incident, the emergency response agency may be in property conservation mode for hours or days. Property conservation, or stopping the loss, is rapidly becoming the fire service's benchmark for excellence.

Relationship between Incident Priorities and Safety

Remember that life safety is always the number one priority and all operations must be developed based on this priority. As an incident priority, life safety applies to both civilians and emergency responders.

Recall from Chapter 3 that risk management began with risk identification. This first step, identification, should be the basis for risk management on the emergency scene. Because the life safety of the responders is the number one priority, incident operations and strategies can be developed that provide the highest level of safety to responders. In other words, for incident stabilization to occur, the incident commander should have considered life safety to the responders in formulating the strategy.

Risk assessment of the emergency scene can be based on the following philosophy:

> We will risk a lot to save a lot,
>
> We will risk little to save a little,
>
> We will risk nothing to save nothing.

Although a risk analysis must always be performed, an incident commander would have few qualms about sending a search and rescue crew into a building fire if there were a possibility that people were inside. However, arriving at the same fire, meeting the occupants outside, and confirming that they are all out may produce a different strategy. Although some argue that a building is never unoccupied until a search proves it, more prudent and safety-minded incident commanders are changing strategies to make them more compatible with the risk.

HAZARDS FACED BY RESPONDERS

In order to do an effective risk assessment and to implement an effective strategy, it is necessary to understand the risks present at fires. For the purpose of this chapter, fire incidents are divided into four groups:

1. Structure fires
2. Transportation fires
3. Outside structure fires
4. Wildfires

Fires related to specialized incidents such as hazardous materials are discussed in Chapter 7.

Structure Fires

Structure fires account for the majority of fires that injure and kill most firefighters. Depending on the makeup of a local community, these may or may not be the largest number of fires responded to. For discussion purposes, structure fires can be further subdivided into one- and two-family dwelling, multifamily dwellings, commercial, and industrial in terms of occupancy. Table 5–1 compares hazards related to each of these occupancies. A similar table could be completed after a risk

Table 5–1   *Hazards associated with various types of occupancies.*

Hazard	Residential 1 and 2 Family	Multistory Residential	Commercial	Industrial
Life hazard	High	High	Medium, depends on time of day	Medium, depends on time of day
Structure failure	X	X	X	X
Large area operations		X	X	X
Exposures threatened by fire		X	X	X
Back draft	X	X	X	X
Flashover	X	X	X	X
Explosions			X	X
Chemical	X	X	X	X
Improper storage	X	X	X	X
Long operations/need for rehabilitation		X	X	X

identification process in your jurisdiction. The hazards listed are general enough to apply to buildings of different size, configuration, and height. However, a number of hazards are common to any structure fire. Some hazards are directly related to the structure in terms of construction. Risk assessment should consider the construction type and the anticipated reaction of the construction methods and materials to fire and heat. For example, it is no secret that by design, lightweight, wood truss roof assemblies have failed early in operations, and the incident commander should consider this in the risk analysis. Hazards relating to the structure that should be considered are:

- Types of loads and transmission of loads (fixed, live, dead, etc.)
- Construction type
- Building material types
- Structural elements including beams, girders, columns, connectors, and support systems
- Avenues for fire and smoke spread

Again, risk identification is necessary to identify local risks.

■ Note
Risk assessment should consider the construction type and the anticipated reaction of the construction methods and materials to fire and heat.

back drafts
occur as a result of burning in an oxygen-starved atmosphere; when air is introduced, the superheated gases ignite with enough force to be considered an explosion

flashover
a sudden, full involvement in flame of materials and gases within a room

mushrooming
heat and gases accumulate at the ceiling or top floor of a multistory building then back down; can be prevented with vertical ventilation

rollover
the rolling of flame under the ceiling as a fire progresses to the flashover stage

heat transfer
the transfer of heat through conduction, convection, radiation, and direct flame contact

Because many structure fires are extinguished from the inside of the building, fire behavior characteristics must also be considered. The incident command and safety manager must be familiar with the signs and potential outcome of situations such as **back drafts, flashovers, mushrooming, rollover,** and **heat transfer** (see Figure 5–4). Through training, education, and experience, the incident commander can make risk assessments and minimize the exposure of responders to these potentially deadly hazards.

Transportation Fires

Fire involving transportation includes vehicle fires, both auto and truck, train, aircraft, and ships. Each of these forms of transportation presents unique problems to the emergency responders (see Figure 5–5). Each form of transportation fire could make a textbook by itself. However, like structure fires, some generalizations can be made and risk identification can assist in assessment of the local risk.

Generally, hazards associated with vehicle fires involve the vehicle itself, the fuel used to power the vehicle, and what cargo the vehicle is carrying. Fire involving the vehicle components themselves may weaken structural components and allow them to fail. When they fail under heat or fire, some vehicle components may explode or become flying missiles. For example, bumpers on vehicles that are designed to withstand low speed collisions without damage work by using a piston. When exposed to heat, the gases in the piston expand and ultimately the piston can fail, sending the bumper hurling toward responders. The fuel that powers a vehicle is a flammable or combustible liquid, which adds

Figure 5–4 *Proper tactics, including ventilation, reduce the chance of back drafts.*

Figure 5–5
Transportation fires can range from the family car to large trucks, railcars, airplanes, or ships.

to the fuel of the fire. And although usually only thought of in the case of trucks, the cargo is also of concern in any vehicle accident. How many people take their small barbeque tank to the refill station in the trunk of their car? Or how many travel home from the pool store with pool chemicals? Vehicle fires large or small require the same risk assessment and application of incident priorities as do structure fires.

Transportation fires most often occur along a roadway so traffic management is also a concern. Because most highway incidents involve vehicle crashes, this subject is discussed in detail in Chapter 6. On-scene risk assessment is necessary to ensure the application of appropriate strategies.

Outside Structure Fires

Fires occurring outside of structures can range from dumpster fires to grass fires. Each can present different levels of hazards and risks. Dumpster or garbage fires require special attention, as the responder normally has little information regarding the contents of the containers. Some clues can present themselves, however. A dumpster behind a hardware store can lead one to assume that paints and other chemicals might be involved. Other possible contacts in dumpster fires include hazardous materials and bloodborne pathogens.

Outside fires also expose responders to certain topographic and environmental hazards, including weather and access hazards. These fires often do not have a civilian life hazard associated with them, so a good risk assessment in these

■ **Note**
Vehicle fires large or small require the same risk assessment and application of incident priorities as do structure fires.

■ **Note**
Dumpster or garbage fires require special attention, as the responder normally has little information regarding the contents of the containers.

situations is risk a little to save a little. This philosophy can go a long way in preventing injuries.

Wildfires

Some of the most publicized fires are wildfires. Whether the fire is in a forest in Wyoming or in the urban interface of California, these fires are large, long in duration, and produce enormous property and environmental loss. They also are responsible for a number of firefighter injuries and deaths each year. Some metropolitan areas never see a wildfire, while for some departments, wildland fires are very common. Because of the nature of these fires, commonly over large geographic areas and for long periods of time, strong incident management systems with good accountability and communication procedures must be in place (see Figure 5–6).

The crews fighting these fires are often very far from help, so it is of critical importance that the incident management system provide for long-term planning and logistical needs, such as food and rehabilitation of crews. Safety program managers and incident commanders who function in wildfire-prevalent areas must have knowledge of weather and the weather's effect on the fire. Included in

Figure 5–6 *Wildfire in progress. Note the onlookers— another safety concern for responders.*

weather are the effects of temperature, air masses, temperature inversions, relative humidity, wind velocity direction, and fuel moisture. For further risk assessment, consideration must be given to topography including elevation and slope, fuel including ground fuels, aerial fuels, structural fuels, and fire behavior. These incidents take the incident management system to high levels, often filling positions that were only designed for special wildland situations. Text Box 5–1 discusses safety tips for wildland firefighting.

Text Box 5–1   *Safety tips for wildland fires.*

Ten Firefighting Orders

1. Keep informed on fire weather conditions and forecasts.
2. Know what the fire is doing at all times.
3. Base all actions on the current and expected behavior of the fire.
4. Plan escape routes for everyone, and make them known.
5. Post a lookout where there is possible danger.
6. Be alert, keep calm, think clearly, act decisively.
7. Maintain communication with your crew, your supervisor, and adjoining forces.
8. Give clear instructions, and be sure they are understood.
9. Maintain control of your crew at all times.
10. Fight fire aggressively, but provide for safety first.

PERSONAL PROTECTIVE EQUIPMENT

Personal protective equipment (PPE) designed for firefighting is one strategy in providing for minimizing exposure at the fire scene and addressing the priority of life safety as applied to responders. PPE for fighting structure and vehicle fires is essentially the same. Wildland firefighting has a different PPE design, although many departments do not provide specialized PPE. Specialized PPE for incidents such as technical rescue and hazards materials is discussed in Chapter 7.

A discussion on PPE for response to fires can be divided into three subject areas:

- Design and purchasing
- Use
- Care and maintenance

Standards from the NFPA set forth requirements for both the purchase and care and maintenance of firefighting PPE. Specifically, *NFPA 1971* includes the requirements for the various components of structural firefighting PPE, while *NFPA 1851* describes the requirements of a PPE care and maintenance program.

NFPA 1500 requires that emergency responders be provided with PPE based on the hazards of the particular work environment. Once provided, guidelines or procedures must be adopted governing the use of the PPE and a care and maintenance program, including inspections.

Design and Purchase

The *NFPA 1500* standard requires that new firefighting PPE meet the current editions of the respective standards. Older gear must have met the standard in effect when purchased. Therefore, the purchasing agent for the department should reference the particular NFPA standard, in its entirety, in the specification for the PPE. The specification writing process, from a safety and health standpoint, should include input from the safety committee, or at a minimum, the safety program manager.

The commonly expected components of PPE for firefighting (see Figure 5–7) include:

- Approved fire helmet with protective eye shield (*NFPA 1971*)
- Flame-resistant hood (*NFPA 1971*)
- Turnout coat (*NFPA 1971*)
- Turnout pants (*NFPA 1971*)
- Firefighting gloves (*NFPA 1971*)
- Firefighting boots (*NFPA 1971*)
- PASS devices (*NFPA 1982*)
- SCBA[1] (*NFPA 1981*)

thermal protective performance rating
the amount of protection against both convective and radiant heat that your composite gear outer shell, thermal liner, and moisture barrier should be able to provide in the event of a flashover

The safety program manager must be involved in the purchasing of firefighting PPE to ensure that the PPE meets the applicable standards. To be effective, the safety program manager must also be familiar with the requirements of the standards and their implications for firefighter safety. For example, the safety program manager must understand the relationship between **thermal protective performance** (TPP) and **total heat loss** (THL). When considered together, these values produce a good indication of the level of safety and performance provided by the PPE ensemble.

Use

total heat loss
the measurement of an ensemble's heat stress reduction capability; the higher the THL value, the greater the benefit for the firefighter

SOPs defining the use of the PPE should also be in place. Although this seems quite obvious, it is helpful from the safety standpoint to require the use of full PPE

---

[1]Although the SCBA unit is usually placed on the apparatus, many departments supply personal face pieces.

Figure 5–7
Firefighters in full NFPA-compliant firefighting PPE.

under specific conditions. Examples might include that full PPE be used for interior structure firefighting. But when is it permissible to remove the SCBA? During overhaul? After the fire is extinguished? The SCBA policy should require the SCBA mask be on until the safety officer or incident commander deems the atmosphere in the work area to be safe. This decision can only be safely made by using air-monitoring equipment. Procedures can also define different levels of PPE for different types of fires. Is SCBA required for wildland fires? How about a dumpster fire? A firefighter would never approach a hazardous materials incident without wearing SCBA, yet would fight a dumpster fire behind a hardware store without SCBA. The use of PPE can be linked to common sense. However, like other components of a complete safety program, the proper use of PPE should be outlined in policies and procedures.

Care and Maintenance of PPE

■ **Note**
The safety program must have a component of PPE inspection procedures, repair procedures, and care.

The NFPA standards on PPE also provide guidelines for the care and maintenance of the equipment. The safety program must have a component of PPE inspection procedures, repair procedures, and care. Generally, the standards require following the manufacturer's recommendations.

Because the PPE is issued to individual members, often the care is also delegated to them. However, because this equipment is specialized, the department has an obligation to provide a process or the facilities for cleaning and minor repairs. In some departments, both of these functions are handled in-house, while other departments contract with outside vendors to provide this service. If the department also has EMS responsibilities, the bloodborne pathogens procedures must also be considered in the cleaning of the PPE.

Although the care is delegated to the user, the safety program must include a procedure for periodic inspection to ensure the PPE is in good condition. Monthly inspections are recommended (see Figure 5–8). The monthly inspection can be performed by a station officer or, in some departments, the shift commander. However, the safety program manager should also be involved in the process. A quarterly or semiannual inspection by the safety program manager or staff is a good idea. After each use, the PPE should be inspected by the user. Of course, SCBA and PASS devices should be checked and inspected at the beginning of each shift and after use. Figure 5–9 provides an example of a PPE inspection form.

Records should be kept on all PPE. Included in the record should be assignment, dates when cleaned, date and type of repairs, and inspection dates. This record should be filed in the workplace of the person who has the PPE assigned to him or her. Table 5–2 is an example of a PPE record.

Figure 5–8 *Company officer doing monthly inspection of PPE.*

EXAMPLE
PERSONAL PROTECTIVE EQUIPMENT RECORD

Employee Name: _____   Year: _____

Item	Inv #	Issued	July		Aug.		Sept.		Oct.		Nov.		Dec.	
Helmet														
Hood														
SCBA Mask														
Mask Bag														
Safety Glasses														
Bunker Coat														
Gloves														
Bunker Pants														
Suspenders														
Bunker Boots														
Work Boots														
HEPA Mask														
/////////			Inp/Res		Inp/Res		Inp/Res		Inp/Res		Inp/Res		Inp/Res	

Inp=Inspection Date and Inspector Initials/ Res=Result of Inspection (Pass or Fail)

REMARKS
July:_____
_____
August: _____
_____
September:_____
_____
October:_____
_____
November:_____
_____
December:_____
_____

Figure 5–9 *A typical inspection check sheet.*

Table 5–2   *Sample PPE record.*

Assigned to: ID#:

Station #:   Shift:

Activity	Helmet	Coat	Pant	Gloves	Hood	Boots	PASS	SCBA	Remarks (Use Back as Needed)
6 Month Clean 1/1/98	X	X	X	X	X	X			
Inspection 1/5/98	X	X	X	X	X	X	X	X	Pass battery due Hole left pant knee
Knee repaired 1/15/98			X						
Battery replace 1/15/98							X		
Annual Flow Test								X	Completed by Smith's SCBA shop

INCIDENT MANAGEMENT SYSTEMS

incident management system (IMS)
an expandable management system for dealing with a myriad of incidents to provide the highest level of accountability and effectiveness; limits span of control and provides a framework of breaking the big job down into manageable tasks

A key requirement to maximize safety of the fire scene is the use of an **incident management system (IMS)**. Further, *NFPA 1561 Standard on Emergency Services Incident Management System* provides the requirments of an incident management system. A number of recognized systems are in use throughout the country. However, as described in Chapter 4, the is a move toward a national interagency IMS as a result of a presidential directive. Local conditions and procedures often dictate the type of system, but to be effective, the system must have the following design characteristics:

- Must provide for different kinds of operations including, single jurisdiction/single agency response, single jurisdiction/multiagency response, and multijurisdiction/multiagency response.

- Organizational structure must be adaptable to any emergency.
- Must be usable and understood by all agencies in a particular geographic area.
- Must be expandable in a logical manner from the beginning of an incident.
- Must use common terminology.
- Must have integrated communications.
- Must maintain a manageable span of control.
- Must be used on all incidents!

Because this is not a text on IMS, the focus is on the role of safety within the IMS structure. The IMS structure provides for a command staff and general staff. Included in the command staff are three functions or positions: information, safety, and liaison. The general staff has four positions: operations, finance, logistics, and planning (see Figure 5–10). Each of these functions is performed at each and every incident. At a single unit response incident, the company officer may fill all the command staff roles; at a one-alarm house fire the shift commander may fill all the roles except for safety and information. At a multialarm commercial fire, administrative officers may fill in many of the command staff or general staff functions. At a several-week-long fire fight at a wildfire, all IMS positions may be filled and the system may be several layers deep.

From a safety and health perspective, the incident safety officer is part of the command staff. To be effective, the safety officer must have expertise in fireground risk management. The safety officer must also have authority within the IMS and direct communication with the incident commander. Requirements and roles of the incident safety officer are further described in Chapter 9.

At an incident scene, the safety officer or member of the safety sector or group must be given certain authorities. These include stopping, altering, and suspending operations that are determined to be unsafe. The safety officer must alert the

■ Note
To be effective, the safety officer must have expertise in fireground risk management.

■ Note
At an incident scene, the safety officer or member of the safety sector or group must be given certain authorities.

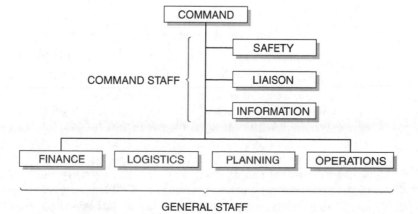

Figure 5–10 *Incident management system—command staff and general staff positions.*

incident commander of the situation and what action the safety officer has taken. The safety officer working at a fire scene should be alert to a number of safety-related actions, proper PPE being worn, structural conditions, establishing collapse danger zones, overseeing the accountability function, preventing **freelancing**, and ensuring that provisions are made for responders' rehabilitation.

freelancing
occurs when responders do not follow the incident plan at a scene and do what they want on their own; a failure to stay with assigned group

ACCOUNTABILITY

Personnel accountability is a critical element of the IMS and of the safety program. Both *NFPA 1500* and *NFPA 1561* require accountability systems at the emergency scene and both place the responsibility for accountability with the incident commander. Accountability occurs by using several layers of supervision in various geographic areas or functional areas. Each supervisor is responsible for his or her operating crews with the concept of company or **crew unity** being of paramount importance. An accountability officer should be assigned within the IMS system, or on smaller incidents, assignments could be handled by the incident commander or safety officer.

crew unity
the concept that a fire company or unit shall remain together in a cohesive, identifiable working group to ensure personnel accountability and the safety of all members; a company officer or unit leader shall be responsible for the adequate supervision, control, communication, and safety of members of the company or unit

As with the IMS, there are several variations in accountability systems in use. Some use the passport system in which the crew's names are placed on small Velcro tags and placed on a small card that is used at the incident scene as a passport (see Figure 5–11). Others use two-dimensional bar codes and computers. Yet another system uses a card and clip that are attached to a large ring carried in the apparatus. Systems can be purchased commercially or designed locally, but the accountability system should meet the following objectives:

- Account for the exact location of all individuals at an emergency scene at any given moment in time
- Provide for expanding to meet the needs of the incident
- Be adaptable to the IMS in use
- Ensure that all individuals are checked into the system at the onset of the incident
- Provide for visual recognition of participation
- Provide for points of entry into the hazard zone
- May also include medical data and the individual's training data

■ Note
Locally adopted accountability procedures are a strict requirement and oblige participation by everyone at all levels and at all incidents.

Locally adopted accountability procedures are a strict requirement and oblige participation by everyone at all levels and at all incidents. For an accountability system to be effective, it must have support by senior management of the organization. Each level has certain responsibilities to the system. Individual responders are responsible to ensure that their name tags are in the proper location on the apparatus and at the scene it is their responsibility to stay in direct contact with their assigned crew. *No freelancing!* Officers are responsible for knowing the location of each person assigned to them and ensuring that they stay within their

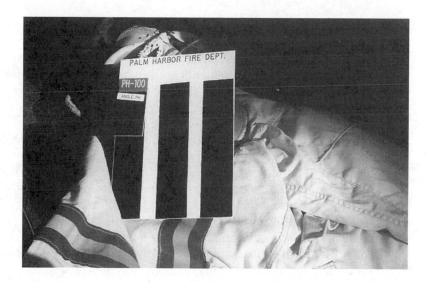

Figure 5–11 *The typical passport and accountability board.*

assigned work area. Sector or group officers are responsible for the accountability within the work area. Accountability officers are responsible for the accountability of any area of the incident from a specific point of entry. Incident command is responsible for putting the accountability function into the IMS and to provide a means for accounting for every individual at the scene. An accountability division/sector/group at a large incident may determine the points of entry, ensure communication with accountability officers, and provide the overall coordination of accountability at a specific incident.

The accountability system requires **personnel accountability reports** (PARs). These PARs require the incident commander or accountability officer to check the status of crews and ensure that all personnel are accounted for during certain regular intervals or benchmarks during the incident. These benchmarks should be defined and may be related to time of activity. Examples of benchmarks include:

- A fixed time, generally every 30 minutes
- After primary search
- After fire is under control
- After a switch in strategy (offensive to defensive)
- A significant event such as a collapse, flashover, back draft
- After any report of a missing firefighter

personnel accountability reports (PARs)

a verbal or visual report to incident command or to the accountability officer regarding the status of operating crews; should occur at specific time intervals or after certain tasks have been completed

RAPID INTERVENTION COMPANIES

Regardless of the effectiveness of the IMS in place, or the greatest of accountability systems, nothing is gained if personnel are not available to respond when an

rapid intervention companies (RIC)
an assignment of a group of rescuers with the sole purpose of rapid deployment to reports of operating personnel in trouble or missing

unexpected event occurs at the emergency scene. The IMS and accountability system can provide the incident commander with the information that a crew is missing, but to intervene, the incident commander must have the resources close by to handle the situation. This is the concept of **Rapid Intervention Companies** (RIC) or Teams (RIT). Although called by many other names, the RIC is a group of firefighters—be it two, three, four, or more—who are fully equipped, have on complete PPE, and who standby near the emergency scene and wait for something to happen. When something does happen—whether it is a lost or trapped firefighter or injured firefighters requiring assistance—this crew is ready to respond and assist. As described in Chapter 1, information from the Phoenix research, clearly revealed that it is necessary for the fire department to practice rapid intervention drills and skills. It was apparent after 200 drills that much more can be learned regarding the use of rapid intervention companies. It is the responsibility of the safety program manager and the training staff of the organization to ensure the rapid intervention company is prepared to be effective, not just an entry on the incident command worksheet.

Although the RIC would meet the requirements of the two-in/two-out rule introduced in Chapter 2, there is ongoing confusion and debate within the fire service as to whether there is a difference—conceptually or functionally—between the two-in/two-out rule and rapid intervention companies. Some fire departments consider the two to be synonymous, that is, the terms can be used interchangeably. Other departments view RICs as a higher order two-in/two-out team that is differentiated by specialized training (Text Box 5–2) and equipment and possibly even the firefighting credentials of RIC members.

For the purposes of this text, the two-in/two-out requirements pertain to initial arrival of units, which is referred to in *NFPA 1710* as the initial RIC. This initial RIC uses the initial arriving units and is deployed as part of first-due operations. Once a scene progresses beyond the incipient stage and escalates to a working fire or additional alarm or both, then the need to deploy a formal RIC or RICs becomes a factor. Firefighter safety is the issue with both RIC and two in/two out. Each has its own requirements and distinct, related approaches to providing safety based on the level of need. Both are implemented and often staffed differently, but are established with the common goal of effecting firefighter rescue.

Rapid intervention crews can be implemented in a number of ways. Some departments dispatch an extra unit to working fires to function as this team; others dispatch a company on the first alarm, for example, the fourth-due engine might function as the RIC. Some use rescue or advanced life support units, others have designated teams such as special operations that respond to all structure fires.

The number of RIC necessary at an emergency incident can vary with the complexity of the incident. A room and contents house fire would probably only require a crew of two or three, while a large warehouse complex may require a RIC on each side of the building simply because of distance.

Text Box 5–2   *Suggested training topics for RIC personnel.* (Courtesy of the USFA/FEMA.)

- Building construction
- Rescue scenarios:
  - Entanglement
  - Floor collapse
  - Confined space
  - Ground level
  - Above ground
  - Below ground
- Incident size-up
- Fire behavior and travel
- Team search techniques and problems (including large-area searches)
- Use of thermal imagers
- SCBA changeover and use of emergency breathing support systems
- Self-rescue techniques
- Fireground communications—frequency and capability
- Forcible exit
- Accountability
- Ladder bail out
- Methods of firefighter removal
  - Unconscious
  - Conscious
- Rope/charged hose line slide
- Command of RIT operations

On arrival, the team should size up the building and the operation and try to anticipate what could happen. It must determine possible entry and exit points. The tools should be assembled for the RIC to use based on the type of building. For example, in a two-story garden apartment building, the RIC may want to have a 24-foot ladder with it for rapid access to the second floor; in a large metal-sided warehouse, the RIC should have a power saw with a metal blade. Text Box 5–3 lists the minimum equipment that should be assembled and carried by the RIC.

This RIC concept has widespread applicability, not only to fire emergencies but other incidents as well. The incident commander must provide for a RIC early in the incident and throughout the duration.

Text Box 5–3 *Minimum equipment for a RIC.* (Adapted from the USFA/FEMA Report.)

- Extra SCBA complete with harness, regulator, and extra masks (consider that automatic/mutual aid companies may use different SCBA systems)
- Search rope
- Forcible entry hand tools such as axe, sledge, halligan bar, and bolt-cutters
- Mechanical forcible entry tools such as chain saw, metal cutting saw, and masonry cutting saw
- Hose line available
- Ladder complement (size and type depends on the building height)
- Thermal imaging camera
- High-intensity handlight

REHABILITATION

The physical and mental demands placed on responders coupled with the environmental dangers of extreme heat and cold will have an adverse effect on the responder from a safety perspective. Figure 5–12 and Figure 5–13 show the heat stress index and the wind chill index, respectively. Crews that have not been provided with adequate rest and **rehabilitation** during a fire or other emergency operation are at increased risk for illness and injury.

Rehabilitation on the emergency scene is an essential element of the IMS (see Figure 5–14). This need for rehabilitation is also cited in a number of the national standards relating to safety and fire scene operations. *NFPA 1584 Recommended Practice on Rehabilitation of Members Operating at Incident Scene Operations and Training Exercises* that applies to personnel operating at emergency scenes and training exercises and requires that the IMS have a component for rehabilitation.

The development of an incident rehabilitation program has minimal impact on an organization and therefore should be achievable. While systems can be very different, the basic plan should provide for the following:

- Establishment of a rehabilitation sector/group within the IMS
- Hydration
- Nourishment
- Rest, recovery
- Medical evaluation
- Relief from climatic conditions
- Accountability while in the rehabilitation sector
- Supplies, shelter, and number of people needed to operate the rehabilitation area

rehabilitation
the group of activities that ensures responders' health and safety at an incident scene; may include rest, medical surveillance, hydration, and nourishment

■ **Note**
Rehabilitation on the emergency scene is an essential element of the IMS.

■ Note

Remember that the medical information from a rehabilitation sector or group may be deemed confidential medical records. Check with your local legal advisor.

Procedures should define the responsibilities within the IMS for the incident commander, supervisors, and personnel. Further, the procedures should provide guidelines for location, site characteristics, site designations, and resources required. Records should be kept on all individuals entering the rehabilitation area and medical evaluation documented. Remember that the medical information from a rehabilitation sector/group may be deemed confidential medical records. Check with your local legal advisor.

RELATIVE HUMIDITY

TEMPERATURE °F	10%	20%	30%	40%	50%	60%	70%	80%	90%
104	98	104	110	120	132				
102	97	101	108	117	125				
100	95	99	105	110	120	132			
98	93	97	101	106	110	125			
96	91	95	98	104	108	120	128		
94	89	93	95	100	105	111	122		
92	87	90	92	96	100	106	115	122	
90	85	88	90	92	96	100	106	114	122
88	82	86	87	89	93	95	100	106	115
86	80	84	85	87	90	92	96	100	109
84	78	81	83	85	86	89	91	95	99
82	77	79	80	81	84	86	89	91	95
80	75	77	78	79	81	83	85	86	89
78	72	75	77	78	79	80	81	83	85
76	70	72	75	76	77	77	77	78	79
74	68	70	73	74	75	75	75	76	77

NOTE: Add 10°F when protective clothing is worn and add 10°F when in direct sunlight.

HUMITURE °F	DANGER CATEGORY	INJURY THREAT
BELOW 60°	NONE	LITTLE OR NO DANGER UNDER NORMAL CIRCUMSTANCES
80°–90°	CAUTION	FATIGUE POSSIBLE IF EXPOSURE IS PROLONGED AND THERE IS PHYSICAL ACTIVITY
90°–105°	EXTREME CAUTION	HEAT CRAMPS AND HEAT EXHAUSTION POSSIBLE IF EXPOSURE IS PROLONGED AND THERE IS PHYSICAL ACTIVITY
105°–130°	DANGER	HEAT CRAMPS OR EXHAUSTION LIKELY, HEAT STROKE POSSIBLE IF EXPOSURE IS PROLONGED AND THERE IS PHYSICAL ACTIVITY
ABOVE 130°	EXTREME DANGER	HEAT STROKE IMMINENT!

Figure 5–12 *Heat stress index.* (Courtesy United States Fire Administration.)

TEMPERATURE °F

WIND SPEED (MPH)	45	40	35	30	25	20	15	10	5	0	−5	−10	−15
5	43	37	32	27	22	16	11	6	0	−5	−10	−15	−21
10	34	28	22	16	10	3	−3	−9	−15	−22	−27	−34	−40
15	29	23	16	9	2	−5	−11	−18	−25	−31	−38	−45	−51
20	26	19	12	4	−3	−10	−17	−24	−31	−39	−46	−53	−60
25	23	16	8	1	−7	−15	−22	−29	−36	−44	−51	−59	−66
30	21	13	6	−2	−10	−18	−25	−33	−41	−49	−56	−64	−71
35	20	12	4	−4	−12	−20	−27	−35	−43	−52	−58	−67	−75
40	19	11	3	−5	−13	−21	−29	−37	−45	−53	−60	−69	−76
45	18	10	2	−6	−14	−22	−30	−38	−46	−54	−62	−70	−78

A   B   C

	WIND CHILL TEMPERATURE °F	DANGER
A	ABOVE 25°F	LITTLE DANGER FOR PROPERLY CLOTHED PERSON
B	25°F / 75°F	INCREASING DANGER, FLESH MAY FREEZE
C	BELOW 75°F	GREAT DANGER, FLESH MAY FREEZE IN 30 SECONDS

Figure 5–13 *Wind chill index.*

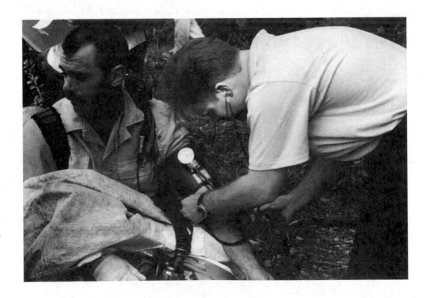

Figure 5–14 *A firefighter beginning the rehab process.*

SUMMARY

Fireground activities account for the highest percentages of firefighter deaths and injuries of any other specific type of duty. It is essential that the safety and health program address safety issues relating to these types of incidents. This requires that the safety program manager understand operational priorities and their application to personnel safety. The operational priorities are always life safety (including the responders), incident stabilization, and property conservation. Understanding the priorities helps to perform risk assessment, which should be based on the premise that we will risk much to save much, we will risk little to save little, and we will risk nothing to save what is already lost.

Hazards faced by responders at fire incidents vary by the type and complexity of the incident. To minimize exposure to the hazards, the safety program must include components relating to personal protective equipment, require the use of an incident management system, provide for an accountability system, ensure that rapid intervention companies are deployed, and have a rehabilitation process. Personal protective equipment concerns are based on design and purchase, use, care, maintenance, and inspections.

The type of incident management system may vary provided that it meets certain criteria. Within the incident management system, there must be a safety officer with the proper authority and responsibility for incident scene safety. Accountability systems provide for a strict accounting of all personnel throughout the incident. When an unexpected situation occurs, the incident commander must have a team available to respond and assist. This role belongs to the rapid intervention company. Prolonged physical and mental stress can adversely affect emergency responders at the incident scene. Proper procedures for rehabilitation must be part of the incident management plan.

Concluding Thought: Having a practical system in place to handle fire-based incidents, including consideration for personnel and equipment, and having the processes in place for safety mitigation will result in safer fireground operations.

REVIEW QUESTIONS

1. Property conservation is the first incident priority.

   A. True

   B. False

2. Explain how the first incident priority can have a positive impact on firefighter safety.

3. List three hazards commonly found at building fires.

4. Which of the following is not an incident priority?

   A. Preplanning

   B. Incident stabilization

   C. Property conservation

   D. Life safety

5. List the ten firefighting orders as related to wildland fires.

6. List the components of an effective PPE ensemble for firefighting.

7. List five characteristics of an effective accountability system.

8. List the positions in the command staff and general staff of an IMS.

9. What three authorities must an incident safety officer be given to be effective on the fireground?

10. How often should firefighting personnel inspect their PPE?

A. Annually

B. Weekly

C. Monthly

D. Daily and after each use

ACTIVITIES

1. Review your department's PPE inspection program. What changes do you feel are necessary to make it more effective?

2. Review your department's procedures for the assignment of rapid intervention companies. Do you use them? Are they effective? What improvements can be made?

3. After performing the risk identification process described in Chapter 3, make a matrix for your jurisdiction relating to risks found at building fires.

4. Review your department's rehabilitation SOP. Does it meet the requirements discussed in the chapter?

Chapter 6

SAFETY AT THE MEDICAL OR RESCUE EMERGENCY

Learning Objectives

Upon completion of this chapter, you should be able to:

- Discuss the hazards faced by responders at emergency medical or rescue incidents.
- Explain the methods of minimizing and preventing injuries associated with hazards found at emergency medical incidents.
- List various requirements and uses of commonly used personal protective equipment.
- Explain the requirements for infection control.
- Discuss procedures that can be used to meet the requirements of infection control.
- Discuss the systems used for scene accountability and incident management.

CASE REVIEW

It was a routine maternity call. The medic unit responded with a crew of two to the apartment and was met by relatives anxious that the day for the newest member of the family to arrive had come. Paramedics went in and found the patient in a back bedroom. The patient was in her early twenties and in her ninth month of a first pregnancy. It seemed routine enough, get vitals, time contractions, start an intravenous (IV) line, and transport. The crew began its work.

As the treatment of the patient began, the crew heard some commotion in the front room. Knowing the large number of family members and their excitement level, they did not think anything of the commo-

tion. Soon a man appeared in the doorway to the bedroom shouting to leave his wife alone. He lurched forward with an 8-inch butcher knife and stabbed one of the crew members in the arm. Police units were summoned and arrived to arrest the man. A second ambulance took the mother-to-be to the hospital. The paramedic who was stabbed was transported by his own medic unit. The injury was minor and, although the medic lost some work, he did recover. It could have been a lot worse.

EMS alarms comprise the majority of alarms that emergency responders face. Although the calls become routine, the potential of risk to responders is high.

INTRODUCTION

Whereas Chapter 5 focused on fire scene and incident safety, this chapter focuses on the type of incidents that most responders face most of the time. It is not uncommon to find rural, suburban, and urban departments involved in EMS, responding to EMS incidents 80% or more of the time. Therefore EMS safety is a critical and integral part of any department's safety and health program.

HAZARDS FACED BY RESPONDERS

What makes the emergency medical incident scene so dangerous? There are many factors to consider when answering this question. One answer might be that, like fire officers, EMS crews must make any number of critical, time-sensitive decisions while on EMS alarms, such as the following:

- The safest and quickest route to the scene
- Drug dosages
- Treatment regimes
- The most appropriate form of transportation and the receiving facility
- Choosing the best interpersonal approach to the patient

Obviously, the decisions occupy a great deal of the responders' thought processes; however, the responder must factor in safety in all of these decisions. A good example might be the response to a multiple car accident on a divided highway that runs from north to south. The accident may be in the northbound lanes, but the ambulance is approaching from the south. The quickest means of reaching the patient would be to drop off a responder and have him or her cross the traffic to reach the victim while the driver went to the next exit and came around. A safer approach, considering, of course, the distance to the exit, would be for both responders to continue to the exit and turn around. This might add an additional few minutes to arrival on the scene, which must be weighed against the threat to the responder in crossing the road. Which approach would you take?

Another question is the validity or accuracy of information from dispatch. Have you ever heard of a response to a man down, which turned out to be a shooting? Not usually the dispatcher's fault, this breakdown of information commonly occurs at the calling party's end. With the increase in cellular phone use, alarms are received sooner, but often the caller is a passerby and does not know a lot about the emergency.

The responders also face safety issues when dealing with the patients, family, and bystanders. Remember EMS events, as with all emergency incidents, are emotional events; people at the scene may not be thinking rationally and they expect immediate action from the responders. Try to tell the frantic father of child who has been electrocuted about the safety issues and the need to cut the power. Or try to tell the outraged family member of the shooting victim that you must wait for the police until you can enter the house.

Like some fire-related incidents, responders to routine emergency medical calls often develop a complacence toward the type of calls (see Figure 6–1). How many chest pains calls have you responded to? Do you approach the patient with

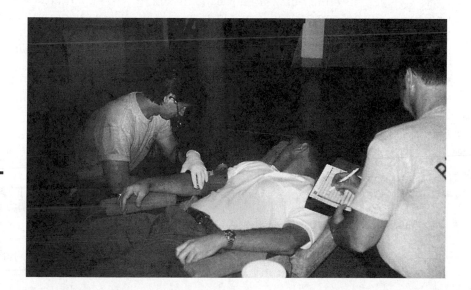

Figure 6–1 *EMS incidents are a safety concern for responders regardless of how routine the incident may seem.*

less caution now than you did your first days on the job? How many helicopter landings? Do you take the same precautions now as you did for your first landing zone setup? The answer very well may be yes, and that is the correct approach. However, the reality is that because you have attended numerous similar incidents and everything has gone well, you may let your guard down. This chapter reminds, reinforces, and, in some cases, gives a new look at EMS incidents and the threats to responders.

General Hazards for All EMS Incidents

Once the responder has arrived on the scene, the incident can be viewed in four phases:

- Gaining access
- Approaching and first contact with the patient
- Providing care
- Packaging and placing the patient in the transport unit

There are obviously additional issues associated with a medical incident, such as response to the scene, but they are addressed elsewhere in this text.

The first phase of the incident is to gain access to the location, whether it be a home, apartment, hotel, restaurant, or vehicle. Upon arriving at the incident scene, park and secure your vehicle in a safe, but accessible, location.

One consideration concerning gaining access are those situations in which forced entry is required. There are a number of concerns, the least of which is security after the patient is removed. Forcible entry is necessary when a patient is unable to come to the door. Law enforcement should be requested to respond to all forcible entry situations and, where possible, the forcible entry should wait until their arrival. Forcible entry is in itself a safety hazard. By definition, forcible entry requires that something will be broken, such as glass or door frames. Responders should have on hand the necessary PPE and the proper tools when attempting forcible entry. If the equipment and tools are not available, then additional resources should be called. Before you force entry, consider that another opening may be unlocked. Many times a second-floor window or balcony door may be left open or unlocked, requiring only the use of a ladder. Before any forced entry occurs, make sure you are at the right location.

■ Note
Before any forced entry occurs, make sure you are at the right location.

In most cases, forced entry is not required and the responder can enter the location as any other person would. But there are still considerations about how you enter and what you see when you do. The responder should always carry a flashlight at night and consider it during the day at some locations. In the event the patient is in a basement or other dimly lit area, the flashlight will be necessary so it is best to carry it always. When looking around on entry, the same considerations should be noted as on arrival outside. Look for access to the patient and consider transport options, poorly repaired stairs, rugs, tripping hazards, and animals. Animals present in the house should prompt the responder to ask that the animal

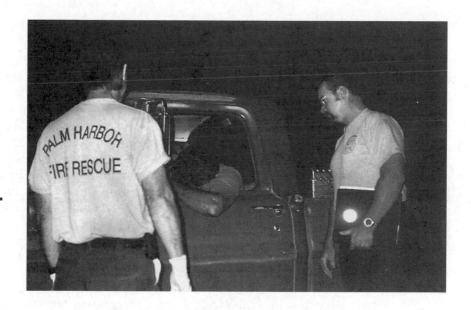

Figure 6–2 *EMS responders should be alert for safety issues when first contact is made with patient.*

be secured in another room. Although the animal may seem friendly or small enough, most animals are very protective of their owners and an attempt to touch or move an owner may induce an unwanted response from the animal.

Once access has been made and the patient located, phase two, approach and first contact, begins (see Figure 6–2). There are a number of things to consider when first contact is made, most of which deal with interpersonal skills. An approach to a conscious patient should start with an introduction and why you are there. If hostility toward the responder exists, it will be apparent at this time. Be aware of threats or statements that indicate that the patient may not want your help. Again this may be an indication that police assistance is needed and a retreat necessary. Note the patient's hands and surroundings and make sure nothing is present that might be used to hurt you or cause injury as you work around the patient. Glass on the floor is a good example, considering you will probably kneel down to take vital signs. Always approach cautiously and move items out of the way that may pose problems in the next phases of the incident.

Unconscious patients should be approached with caution as well. Is it possible that you might be waking up a patient who just paid $100 for a high and passed out? Or is it possible that the patient is just sleeping and could react violently when awakened? Again, a cautious approach and being prepared to back off is required.

During first contact, the responder must use a variety of interpersonal skills. Tone of voice is important: Commanding respect would be necessary in a nightclub, while soft and gentle might be the order for a small child at home. An understanding of personal space is important, remembering that in the space in which the responder works, taking vital signs, holding spinal support, and the like

■ Note
During first contact, the responder must use a variety of interpersonal skills.

involve touching the patient. It is generally accepted in the United States that intimate space is 18 inches around a person. Therefore, any unfamiliar responder immediately invades this space and the patient may experience a feeling of uneasiness, leading to an unwanted reaction. It is, therefore, very important for the responder to develop trust early on.

After, and almost simultaneous with, first contact is the providing of care phase of the response. In this phase the responder is again faced with several safety challenges, one of which is exposure to communicable diseases, which is discussed under Infection Control. Others include injury during moving the patient, needle sticks, and contact with objects around the patient. Usually if the necessary precautions have been taken in the first two phases and the situation does not involve communicable diseases or hostility, this phase can proceed relatively safely.

The final phase of the EMS incident to be discussed in this chapter is packaging and placing the patient in the transport unit. Now is the time when many of the mental notes that were made on arrival become important. Are the stairs safe to carry the patient? If not, what other means are available? Consider that stairs not only have to support the weight of the responders, but also the patient being carried. Perhaps a fire department ladder truck with a bucket might be a safer alternative. Were there any obstructions or tripping hazards, either inside or outside, on the approach? Are rugs and carpets secured down? If not, roll them up to create a clear path. Finally, are enough people present to make the move safely, considering outside weather conditions, the patient's weight and location, and available equipment.

Hazards Associated with Vehicle Accidents and Incidents on Roadways

Operations associated with incidents on roadways pose distinct safety hazards, as compared to incidents inside a structure. Responders to incidents on roadways have a great deal less control over their surroundings (see Figure 6–3). Traffic, weather, lighting conditions, and crowd control are concerns. Responders' safety on the highways has started to receive a great deal of attention, and rightly so.

On arrival at this type of scene, the first order of tactical concern is where to safely locate the response vehicle. The first-arriving unit must be placed to have an initial blocking of oncoming traffic in an effort to create a safe work area. Consideration should be given to parking the unit in a manner that will provide the best level of protection for the responders. Other considerations are which location will best facilitate patient care and transport, which location will provide best departure, and whether other emergency vehicles will be arriving. Of course, safety is the primary concern when considering vehicle placement. Larger vehicles, such as fire apparatus, should be positioned to block the scene; smaller vehicles such as ambulances and law enforcement vehicles should be placed inside the protection of the larger vehicles. Locating the unit also depends on other hazards that

Figure 6–3 *For the safety of the responders, incident commanders must consider stopping all traffic on the roadway while fire and EMS operations are underway.* (Photo courtesy of Monroeville VFC #4.)

could be present at the scene, such as fuel spillage, downed power lines, or hazardous cargo. Always avoid parking the vehicle in a manner that requires emergency responders to cross traffic to get to and from the scene.

Traffic always presents hazards to responders to roadway incidents. Several safety principles should be applied. Create a buffer zone by using an emergency vehicle as described in the previous paragraph. Patient loading areas must be within the protected area afforded by the larger apparatus. Do not be afraid to close the road completely to protect responders' safety. This can be accomplished by using orange traffic cones and law enforcement personnel. Finally, wear protective clothing or a vest with reflective trim regardless of whether the incident is in the daytime or after dark (See Figure 6–4). Text Box 6–1 provides some general guidelines for personnel operating on highways.

Other hazards related to roadway incidents, specifically vehicle accidents, are those hazards associated with vehicle rescue or extrication. **Vehicle rescue** is comprised of action to gain access, disentangle, and remove persons trapped in vehicles. Most often this is accomplished using heavy hydraulic equipment for cutting and spreading metal and car components (see Figure 6–5). Aside from the

vehicle rescue

comprised of action to gain access, disentangle, and remove persons trapped in vehicles

Figure 6–4
Responders must wear protective clothing or a vest with reflective trim regardless of whether the incident is in the daytime or after dark. (Photo courtesy of Firefighter/ Paramedic Chris Eisenhardt.)

Text Box 6–1   *General safety principles for responders operating at highway incidents.*

- Wear proper PPE including reflective vest and helmet.
- Maintain an awareness of traffic at all times.
- When personnel permits, have a spotter watching oncoming traffic at all times.
- Have a system to alert responders on the scene of a threat. This system might be an air horn on the fire apparatus similar to a fire scene evacuation signal.
- Avoid turning your back to oncoming traffic.
- Check traffic before you move.
- Look before exiting the response vehicle.
- If department policy, deploy cones and at night, flares. Face traffic when deploying these devices. Be aware of flare use or smoking by bystanders around leaking fluids from the vehicles.

hazards of the accident itself, using this equipment produces the following additional hazards:

- Flying glass and metal
- Potential of rescuers' fingers and hands being caught between the vehicle and the cutting or spreading tool

Figure 6–5   *Examples of heavy extrication equipment.*

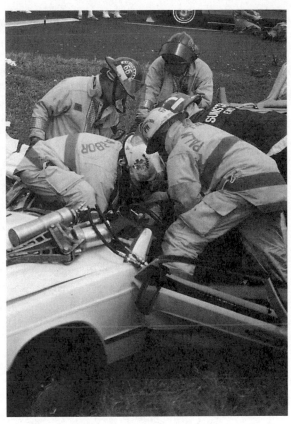

Figure 6–6   *A vehicle extrication in progress.*

- Fire hazards as a result of using powered tools
- Injuries resulting from lifting heavy tools and equipment

Any time vehicle rescue operations are being performed, an IMS should be established, a safety officer assigned, full protective clothing designed for thermal and abrasion resistance should be worn, and fire suppression capabilities should be present (see Figure 6–6).

In addition to vehicle accidents, other EMS-needed incidents occur on roadways. Injuries from violence, including shooting, stabbing, and assault, are too prevalent in the country today. Responders to these incidents should be aware of these situations and rely on law enforcement support and scene control. The hazards associated with these incidents are obvious. At a drive-by shooting, for example, who is to say the shooters will not drive by again after responders have arrived.

Environmental hazards are also a concern in roadway situations. Extreme cold or hot, rain, snow, sleet, and the duration of the operation can impact the responders' safety. Rehabilitation considerations similar to those presented in Chapter 5 should be a part of the response plan.

Chapter 4 discussed the importance of interagency preplanning and interaction. This is very important for roadway incidents. Law enforcement is needed for crowd and traffic control and to ensure the securing of the scene following incidents involving violence. Preplanning the necessary equipment, resources, and expectations of both agencies is essential to maximize safety on the scene.

INFECTION CONTROL

Infection control is a requirement for emergency responders to EMS incidents. Chapter 2 presented federal regulations and an NFPA standard that govern these programs. Specifically OSHA 1910.1030 deals with bloodborne pathogens and *NFPA 1581* gives the requirements for a fire department infection control program. Infection control is essential, not only for responders' safety, but also to maximize the safety of the patients they treat. The safety program manager may or may not be the department's infection control officer. Regardless of who it is, the law requires the organization assign an infection control officer to handle exposures, follow-up care, and record keeping. The organization must also adopt an infection control plan to prevent the transmission of diseases.

The infection control plan is based on engineering controls, personal protective equipment, and education. Engineering controls can be based on risk management principles. For example, the risk of needle sticks can be reduced or eliminated by changing the practice of resheathing needles and just placing them in a sharps container. Personal protective clothing is discussed in the next section. Education requires training employees in the types of infectious diseases, routes of transmission, and prevention strategies.

■ Note

Responders are exposed to a number of diseases, some minor, some serious, some even fatal.

Responders are exposed to a number of diseases, some minor, some serious, some even fatal. An exposure is not the single mechanism needed to acquire the illness; several factors must interplay for an infection to occur. As with injury prevention, the infection can be prevented by interrupting the disease process at any of these points:

Dose: the number of live organisms present

Virulence: the strength of the infecting organism

Host Resistance: the ability of the host to resist the effects of the infectious organisms

Route of Exposure: airborne, bloodborne, or foodborne

Means of Transmission: a way for the organism to gain entry into the host—injection, inhalation, absorption, or ingestion

An infection control manual can provide information regarding the listed factors for specific diseases. One of the best ways to intervene or interrupt the disease process is to prevent the means of transmission. Diseases are transmitted via direct or indirect contact, an intermediate host, or other vehicles.

Direct contact transmission occurs when physical contact occurs between the responders and the infected person. Indirect contact occurs when an infected person contacts an object and then the responder contacts the same object. Of course, in the case of bloodborne disease, there would have to be blood on the object. Another means of indirect contact is droplet contact, which occurs when an infected person coughs or sneezes and airborne droplets are produced. These microscopic droplets can travel a great distance.

Examples of transmission via an intermediate host would be diseases spread by ticks, mosquitoes, flies, or fleas. Disease spread through water and food are examples of transmission via other vehicles.

As described in Chapter 4 in the section on employee wellness and fitness, responders should take advantage of and be provided with boosters and vaccinations for diseases when available. Responders should also be trained in personal hygiene and equipment cleanup procedures. The response station must have clearly marked areas for cleaning contaminated equipment. In terms of personal hygiene, hand washing is particularly important. Since hand-washing facilities are not always available, waterless gels that are disinfectants should be carried on the emergency vehicle.

■ **Note**
Responders should take advantage of vaccinations and boosters for diseases when available.

PERSONAL PROTECTIVE EQUIPMENT

PPE for responders to EMS incidents is specialized and differs from that of firefighting PPE. EMS responders who are cross trained as firefighters are issued firefighting PPE that is appropriate for most EMS incidents, such as vehicle rescue, because firefighting PPE provides thermal and abrasion resistance. Because firefighting PPE was described in Chapter 5, this section focuses on PPE particular to EMS emergencies that are unrelated to entrapment or rescue.

PPE requirements for **body substance isolation (BSI)** are defined in the respective regulations and standards. The EMS providers often must make a decision as to what PPE is necessary at particular incidents. **Universal precautions** should always be used any time there is a chance to come into contact with potential infected body fluids or materials with body fluids on them. Procedures should be developed to give guidance in PPE selection for tasks at the emergency scene. Table 6–1 is an example of a matrix that should be included as part of an infection control SOP. It is *always* the rescuer's option to do more than the minimum requires based on the circumstance. From head to toe, the EMS responder should have PPE adequate for the hazards to be faced. Table 6–2 charts PPE for various body areas. These are recommendations and must be geared to local conditions.

EMS PPE should be cared for, disposed of, and inspected according to the manufacturer's recommendations. The inspection program should be similar to

body substance isolation (BSI)
an infection control strategy that considers all body substances potentially infectious and isolates all body substances that might be infectious

universal precautions
term used to describe the practice of treating all patients as if they carried an infectious disease, in terms of body fluid contamination

Table 6–1 *PPE matrix.*

Task or Activity	Disposable Gloves	Gown	Mask	Protective Eyewear
Bleeding control with spurting blood	X	X	X	X
Bleeding control with minimal bleeding	X			
Emergency childbirth	X	X	X	X
Blood drawing	X			X
Starting an intravenous (IV) line	X			X
Endotracheal intubation, esophageal obturator use	X		X	X
Oral/nasal suctioning, manually cleaning airway	X		X	X
Handling and cleaning instruments with possible microbial contamination	X	X	X	X
Measuring blood pressure	X			
Giving an injection	X			
Measuring temperature				
Cleaning back of ambulance after a medical alarm	X	X	X	X

that of fire PPE, however, much of the EMS PPE is disposable. The same record keeping should be required for EMS PPE as for firefighting PPE.

INCIDENT MANAGEMENT SYSTEMS

Like fire incidents, emergency medical/rescue incidents must be managed with an IMS. This is not to say that responders must break out the command board for every heart attack call, but in reality IMS does occur on each incident. It is just that the senior EMS responder assumes many of the roles in the command staff. On multiple unit or multiple jurisdictional response, IMS is as much required on the EMS incident as it is on the fire or hazardous materials scene.

For an EMS incident, the IMS can be adapted. The three command staff and four general staff functions remain the same. Responsibilities within the command

Table 6–2 *Personal protective equipment for body areas.*

Body Area	PPE Item	Remarks
Head	Helmet	If not provided as firefighting PPE
Head	Eye protection	To protect from objects and body fluid intrusion
Head	Hearing protection	For prolonged rescue in noisy environment
Head	Mask	To prevent airborne transmission or fluid from entering the mouth
Torso	Abrasion-resistant jacket	If not provided as firefighting PPE
Torso	Gown	To prevent fluid transmission
Arms	Protective sleeves	To prevent fluid transmission
Hands	Abrasion-resistant gloves	If not provided as firefighting PPE
Hands	Fluid-resistant gloves	
Legs	Abrasion-resistant pants	If not provided as firefighting PPE
Legs	Gown	To prevent fluid transmission
Feet	Shoes hard sole, steel toe	
Feet	Shoe covering	To resist fluid penetration
Body	Body armor	Dependent on local conditions and risks

■ Note

On a routine medical incident, the senior person or officer must assume all of the command functions including safety.

structure only differ depending on the type of incident. At a multicasualty bus accident without entrapment, there may be a group/sector for triage, treatment, and transport. At a single vehicle in a pole with entrapment, there may be groups or sectors for hazard control, extrication, and treatment. The system is adaptable to any emergency. One common factor is the incident safety officer. Although very dependent on the situation, the safety officer should be assigned any time responders are performing hazardous operations such as extrication. On a routine medical incident the senior person or officer must assume all of the command functions including safety. In this role, the safety officer can ensure and require that proper PPE is being worn for the particular type of incident. Figure 6–7 details an expanded IMS at a medical/rescue incident.

An accountability system for EMS responders must be adhered to as with fire incidents. In some jurisdictions, some parts of EMS response are handled by private or third service public agencies. When these agencies respond and are required to work at the incident they must be part of and work within the IMS structure and the accountability system. The safety program manager for the EMS provider should ensure that all EMS personnel have the proper training and equipment to work within these two systems. This also can be part of interagency coordination.

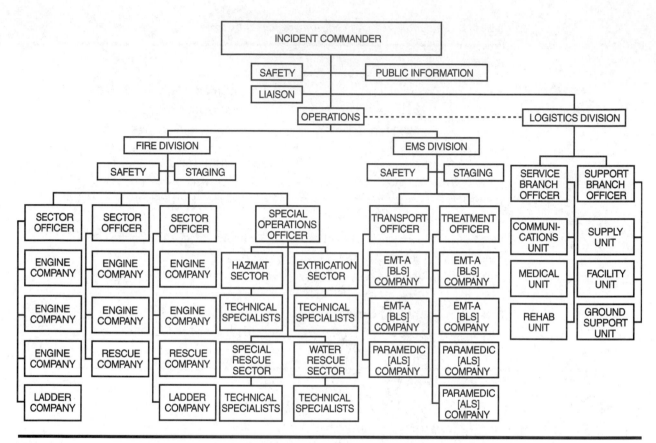

Figure 6–7 *An example of an expanded IMS including EMS functions.* Note: *The EMS division is operational; the medical unit is logistical.* (Courtesy United States Fire Administration, Emmitsburg, MD.)

SUMMARY

For most jurisdictions, EMS emergencies comprise the highest majority of incidents. The hazards faced by responders are increased by uncertainties and the need to interact closely with the patients. Medical or rescue incidents can be generally divided into those that occur inside and those that occur outside. Inside structure incidents, often viewed as routine, can cause injuries. A step-by-step approach and an awareness of surroundings can reduce this potential.

Outside incidents usually involve vehicle accidents. Hazards such as the vehicle, the fuel for the vehicle, traffic, and the environment cause additional problems for the responders. Anticipating these hazards through training on proper vehicle placement and control of the scene can reduce exposure to these hazards. Utilizing law enforcement for crowd and traffic control is also required.

Infection control is an additional concern in medical response. Through the use of engineering, education, and PPE, the risk of acquiring an infectious disease can be reduced. The use of PPE is both required and an excellent means to interrupt the disease transmission process.

The IMS must be adaptable to EMS events and provide for the same command functions and structure as that of fires. The system must include a method for scene accountability of EMS responders, particularly if the EMS responders are from a different agency.

Concluding Thought: EMS and fire events have many of the same requirements for success. Engineering, education, and PPE are three requirements essential to the safety program.

REVIEW QUESTIONS

1. List four hazards associated with EMS emergencies that occur indoors.

2. List three hazards that would be present outside on a highway at a vehicle accident.

3. Which of the following practices should a responder to a vehicle accident *not* take?

   A. Create a buffer zone

   B. Create a nontraffic lane

   C. Park so that traffic must be crossed to access patients

   D. Wear reflective vests

4. What three functions might be in the IMS for a multicasualty incident?

5. List the five factors that must interplay for an infection to occur.

6. Which of the factors in Question 5 is easiest to control?

7. In terms of the spread of infection, compare the terms *direct* and *indirect contact*.

8. What are the three components involved in vehicle rescue?

9. What PPE components might also be part of firefighting PPE?

10. List five tasks or activities that should be included in a required PPE matrix.

ACTIVITIES

1. Review your department's procedures and practices for responding to vehicle accidents. Do these apply the safety principles as discussed in this chapter?

2. Review your department's infection control plan and compare it to the federal requirements. Do you have an infection control officer, and does this officer regularly respond to incidents? Does the plan provide for engineering controls, education, and appropriate levels of PPE?

Chapter 7

SAFETY AT SPECIALIZED INCIDENTS

Learning Objectives

Upon completion of this chapter, you should be able to:

- Describe the safety issues related to hazardous materials incident response.
- Describe the safety issues relating to technical rescue operations.
- Explain safe procedures to be used during helicopter landing zone operations.
- List specific safety issues relating to operations at civil disturbances.
- List specific safety concerns when responding to terrorism events.
- List specific safety concerns when responding to natural disasters.

CASE REVIEW

One relatively quiet Sunday afternoon, the crew of a medium-sized Florida fire department received a telephone call from a concerned emergency room nurse from a nearby hospital. She reported that a patient had been brought into the emergency room via a private car. The patient complained of shortness of breath, but the cause was unknown. As the patient was elderly, normal care for shortness of breath was established. During the routine physical exam, burns that appeared to be chemical burns were noted in the patient's mouth. With the patient's level of consciousness in question, the nurse called the fire department to ask about possible hazardous materials incidents that day. The fire department had not had any hazmat-related calls that day, but took down the patient's address and assured the nurse that they would look into it.

The address was in a local mobile home park. A single engine and battalion chief responded to check the address. Prior to arrival, the crew discussed an approach strategy. Because the potential existed for some type of chemical at the location, a very cautious approach was undertaken. Upon arrival, units staged upward, donned protective clothing, and, with monitor in hand, approached the address. At the front door the vapor monitor went into alarm mode and the crews backed off, setting up a perimeter, and evacuating several mobile homes nearby.

The fire department discovered a lethal amount of ammonia in the mobile home. The incident lasted about 16 hours until the ammonia leak could be bled off. The outcome was good; the patient did well, and the responders operated in a safe fashion.

But, what if the patient had remained at home and called 911 and reported himself to be short of breath. How would the crew have responded? Probably the same way they had responded to hundreds of other shortness of breath incidents. However, in this case they may have entered a mobile home and been exposed to high amounts of a hazardous substance. Responders must always be aware of potential hazards, even when responding to seemingly routine incidents.

INTRODUCTION

The discussion of safety issues at emergency scenes would not be complete without a look at safety and special operations. These particular types of incidents include hazardous materials response, technical rescue incidents, helicopter landing zone operations, operations at civil disturbances, terrorism events, and natural disasters.

These specialized incidents are addressed in a separate chapter because some departments may not be involved in all of these operations. Some departments may not be involved in any of them. The local conditions and the services that a particular department provides dictate what safety measures and procedures must be in place. Each type of incident is discussed and basic safety concerns presented.

HAZARDOUS MATERIALS INCIDENTS

Hazardous materials response is routine to many departments. What differs significantly is the level of response. Some departments may have full hazardous materials teams and provide complete response services including offensive action to mitigate the incident. Other departments may respond at the first response level and provide defensive operations only, relying on a regional hazardous materials team for offensive action and mitigation. OSHA 1910.120 and EPA regulation 311 govern the response to hazardous materials incidents. There are also two NFPA standards and one Recommended Practice applicable to hazardous materials:

- *NFPA 471 Recommended Practice for Responding to Hazardous Materials Incidents*
- *NFPA 472 Standard for Professional Competence of Responders to Hazardous Materials Incidents*
- *NFPA 473 Standard for Competencies for EMS Personnel Responding to Hazardous Materials Incidents*

These regulations require that certain safety measures be in effect before operations can begin. These requirements are described in Chapter 2 (see Figure 7–1).

As part of the overall safety and health program, the safety program manager must determine what level of response the department provides and compare that level to the training level of the responders. There are five levels of response training, as shown in Text Box 7–1.

■ Note

As part of the overall safety and health program, the safety program manager must determine what level of response the department provides and compare that level to the training level of the responders.

Figure 7–1
Hazardous materials response and operations produce unique safety concerns.

Text Box 7–1   *Five levels of hazmat response training.*

First Responder Awareness: Responders at this level are likely to witness or discover a hazardous substance release and have been trained to initiate an emergency response sequence. At this level of training, these responders would be expected to take no further action. Typically, single-certified EMS responders and law enforcement officers fall into this category.

First Responder Operational: Responders trained to this level respond to hazardous materials incidents as part of the initial response for the purpose of protecting nearby persons, property, and the environment from the release. These responders are trained in defensive tactics without actually trying to stop the release. Their function is to contain the release from a safe distance, keep it from spreading, and protect exposures. First responder operational level requires at least 8 hours of training or demonstrated competency.

Hazardous Materials Technician: Responders trained to this level assume a more aggressive role than a first responder operational level. The hazardous materials technician approaches the point of release in order to plug, patch, or otherwise stop the release of a hazardous substance. Hazardous materials technicians must have received at least 24 hours of training equal to the first responder operational level and demonstrated competency in additional areas.

Hazardous Materials Specialist: Hazardous materials specialists are those individuals who respond to a hazardous materials scene to support technicians. Their duties parallel those of the hazardous material technicians, but they require a more direct knowledge of the various substances they need to contain. Hazardous materials specialists may also act as the liaison with federal, state, local, and other governmental authorities in regard to site activities. Hazardous materials specialists must have received at least 24 hours of training equal to the technician level, plus additional competencies as identified in the standard.

On-Scene Incident Commander: Incident commanders who will assume control of the incident beyond the first responder awareness level must receive at least 24 hours of training equal to the first responders operational level and, in addition, have competency in command systems, response plans and options, and other hazards associated with hazardous materials.

Having identified the five levels of training, the program safety manager must review department procedures and practices to ensure that the responders are operating within the level in which they have been trained and within the level of equipment available. A serious incongruence would occur if the department responded routinely to hazardous materials releases and plugged the leaks, but had only been trained to a first responder level.

There are numerous safety-related issues regarding the hazardous materials incident. The risk factors increase with the level of service provided. Hazardous materials PPE is a complete subject in itself. Regulations require the use of incident command systems and backup teams with the same level of protection as the entry team. However, many of these issues relate only to departments that provide hazardous materials technicians and do offensive mitigation strategies. These safety measures should be addressed in the hazardous materials teams operating procedures.

One area of hazardous materials response that is applicable to all responders is the initial response and arrival. Procedures and training for this segment of the incident should focus on safety and the initial response. One method to use to maximize safety of the initial response is to remember the acronym RAID, whose letters stand for Recognize, Approach, Identify, and Decide. See Text Box 7–2.

Text Box 7–2 *The RAID process.*

Recognize. The recognition phase is when the responder recognizes that an incident might involve a hazardous substance. The recognition process begins with preplanning. Being aware through preplanning what hazards are present in a given response area helps to determine the precautions that will be taken at an incident at that specific location. Clues to the presence of hazardous materials also present themselves at the time of dispatch: The location, occupancy, and the name of the business can provide clues. For example, a report of a person(s) down behind Jones Pool Chemicals should give the responders a pretty strong clue that hazardous materials may be involved.

Approach. During the approach phase, several things must be considered. If the responders suspect that a hazardous substance is involved, the approach should be cautious and from uphill and upwind if possible. During the approach, the responders should be looking for other clues, such as strange-colored smoke, vapor clouds, and employees running from the area. If necessary, park the response unit away from the scene until further analysis can be performed. In the Jones Pool Chemicals example, the crew should be alert for vapor clouds or smells such as chlorine fumes.

Identify. Once the scene has been safely approached by responders and if there is a strong suspicion of a hazardous substance incident, they may alert the hazardous materials response team. However, if further information is required, the first responder can start the identification process. Text Box 7–3 outlines some of the resources available for hazardous substance identification.

Decide. The final step in the RAID process is decide. At this point, the responders should have a reasonably good idea if they are equipped and trained to proceed and take action. If the hazard exceeds the capabilities of the responders, specialized teams should be called.

Awareness of safety issues at a hazardous substance emergency is important for all responders. Individual departments must access their level of response and develop appropriate safety procedures.

Text Box 7–3   *Some resources available for hazardous substance identification.*

Occupancy and Location. The mere location of an incident or the occupancy of the building or vehicle may provide clues to identification.

Placards and Labels. Hazardous substances in transit will be placarded according to U.S. Department of Transportation guidelines (see Figure 7–2). However, if the load is less than 1,000 pounds, some materials are not required to have a placard. If the placard contains the four-digit identification number, the product can be identified. If not, then just a generalization as to the type of material is provided, for example, a flammable gas. Labels are similar to the placards but are found on the individual packages.

Shipping Papers. Regardless of the mode of transportation, some type of shipping papers should be with the operator; therefore in a train they are in the engine or in a truck they are the cab. These papers provide identification of the product as well as hazards.

NFPA 704 System Placard. Fixed facilities may have an *NFPA 704* placard on the area where hazardous substances are stored or used. This placard is a diamond divided into four small diamonds. Each diamond is color coded for the hazard and assigned a number from 0 to 4, with 0 being no hazard and 4 being an extreme hazard. The color code is red for flammability, blue for health, yellow for stability, and the white diamond for special information.

Material Safety Data Sheets. Fixed locations storing or using hazardous substances are required to maintain material safety data sheets, which provide product identification information, hazards, and procedures for emergencies.

Employees and Occupants. Employees working at the incident location may also be able to provide product identification information. Remember, however, that employees often downplay the event. Information should be backed up with reference information. Certain facilities may also have internal fire brigades.

Figure 7–2 *Placards and labels help responders identify products.*

ChemTrec. 1-800-424-9300 may be called for assistance in chemical emergencies not only to assist in identification but also for emergency action. ChemTrec can usually put the caller in touch with product experts or shippers.

Reference Books. Although a number of reference books exist, no emergency response unit should be without, at a minimum, the Department of Transportation's *Hazardous Materials Guidebook*, the orange book.

Symbols. Symbols may provide at least general information regarding the hazards of the chemicals. Together with the specific information regarding the physical and health hazards of the hazardous chemical, symbols can be useful in the identification process.

TECHNICAL RESCUES

NFPA 1670 is a standard that addresses technical rescue operations and includes the following categories of technical rescue:

- Structural collapse
- Trench and excavation
- Rope rescue
- Confined space rescue
- Rescue from vehicle and machinery
- Wilderness rescue
- Rescue from water

Each of these types of incidents requires a great deal of expertise and specialized equipment to handle. Much like the hazardous material responders training levels, the NFPA has developed training levels for technical rescue responders. These training levels are described in Text Box 7–4.

Text Box 7–4   *Technical rescue responder training levels as defined in* NFPA 1670.

Awareness. The awareness level represents the minimum capability of responders who as part of their normal job duties could respond to, or be the first on the scene of, a technical rescue incident. This level can involve search, rescue, and recovery operations. Members of a team at this level are generally not considered rescuers.

Operations. The operations level represents the capability of the responder to perform hazard recognition, proper equipment use, and perform the techniques necessary to safely and effectively support and participate in a technical rescue incident. This level can involve search, rescue, and recovery operations, but usually operations are carried out under the supervision of technician-level personnel.

Technician. The technician level represents the capability of the responder for hazard recognition, proper equipment use, and performance of techniques necessary to safely and effectively coordinate, perform, and supervise a technical rescue incident. This level can involve search, rescue, and recovery operations.

These types of incidents present a number of safety issues to the responder. The safety program manager along with the organization's management must determine the level of response to be provided and equip and train personnel accordingly (see Figure 7–3).

Each technical rescue response presents different safety issues from air supply in confined space to the threat of secondary collapse in structural collapse rescue. Generally, technical rescue incidents last longer than fire and emergency medical alarms. Rehabilitation is of prime concern during these incidents. Technical rescue incidents must be managed with an IMS and a safety officer in place.

■ Note
Rehabilitation is of prime concern during these incidents.

Figure 7–3
Technical rescue operations require specialized training and equipment. (Photo courtesy of Pinellas county EMS and Fire Administration.)

The safety officer should be familiar with the operation in question. General safety issues relating to technical rescue incidents include:

- Environmental conditions. Responders will be exposed for long periods of time.
- Stability of the building, trench, or confined space.
- Duration of available air supply, if necessary.
- Resources of personnel and equipment.
- Safe atmosphere—flammable gases? oxygen deficient?
- Rehabilitation of responders.
- Backup teams similarly trained and equipped.
- Logistics, food, rest, and so forth for long-term operations.

Each technical rescue incident must be evaluated for risk versus gain. Often these incidents are actually body recovery situations; responders' safety must not be compromised. Technical rescue is a relatively new discipline for the emergency responder and is evolving. Suggestions for the incident safety officer assigned to a technical rescue incident are provided in Chapter 9.

HELICOPTER OPERATIONS

Emergency responders are often called on to perform operations dealing with helicopters. These operations may include working with the helicopter during a rescue, deploying responders from the helicopter into water, setting up a landing zone for a medivac helicopter, or providing ground support for a helicopter fighting a wildfire. Clearly, some of the incidents are specialized and require a great deal of training and coordination with the helicopter crew; however, some general safety precautions, can be taken when operating around helicopters. Primarily this section deals with ground support operations and approaching the aircraft. The most common ground support operation involving helicopters is the establishment of landing zones. Several safety principles present during this function.

■ Note
The landing zone should be free from flying debris. Remember, if the aircraft weighs 2,000 pounds, then 2,000 pounds of air must be displaced for the aircraft to fly.

Landing Zone

Be sure to select a proper landing zone. This zone will differ with jurisdiction and between day and night operations. Generally speaking, the size of the landing zone depends on the type and size of the aircraft. Landing zones can range from 60 to 120 feet square. The landing zone should be on as level ground as possible. A pitch of greater than 8 degrees to 10 degrees is risky. The landing zone should be free from flying debris. Remember, if the aircraft weighs 2,000 pounds, then 2,000 pounds of air must be displaced for the aircraft to fly. This produces very high wind on landing and takeoff. Garbage can lids and other debris can be picked up and moved about in this wind, possibly even into the blades.

Marking the landing zone is based on local procedures. Many jurisdictions require five markers, one on each corner of the landing area and the fifth to be placed on the windward leg in the center to show wind direction. At night, the landing zone markers must be lit. Usually the marker lights are blue. Some jurisdictions mark the landing zone with headlights of vehicles. Although this practice is acceptable, it must be cautioned that the lights must stay aimed so as to not shine up into the pilot's eyes. It is not good practice to use flares or cones.

Fire suppression must also be available at the landing zone. The fire suppression unit should be specifically assigned to landing zone duties and an IMS should be in place. The crew should have full protective clothing on and be prepared to deploy a hose line for fire suppression. The landing zone fire suppression unit should be parked so as to not be a part of an emergency should one occur. Responders standing by or working near the helicopter should be provided with eye and hearing protection in addition to firefighting PPE. Figure 7–4 shows a typical landing zone configuration for a night operation.

Crowd Control

The landing of a helicopter often draws many onlookers. Vehicle and crowd control are an important function around the landing zone. Often a helicopter will

■ Note

Responders standing by or working near the helicopter should be provided with eye and hearing protection in addition to firefighting PPE.

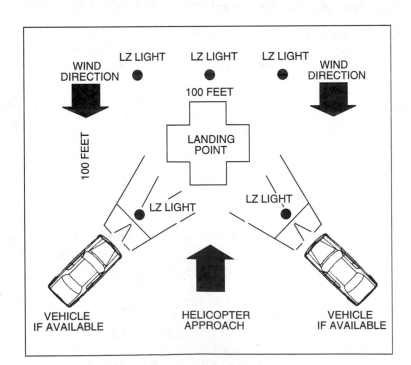

Figure 7–4 *Typical landing zone configurations for night operations.*

land and keep the motors running so there is no delay in departure. Nonessential people, including bystanders and responders should be kept a safe distance from the aircraft. Although this crowd and traffic control is, in most cases, a police function, other emergency responders may find themselves involved. It is a good idea to keep bystanders and nonessential emergency personnel at least 200 feet from the aircraft.

Approaching the Aircraft

In many cases emergency responders will, at some point, have to approach the aircraft, whether it be to load a patient or hook up a hose line to refill a drop tank. The safety principle to be applied in this case is to *never* approach the aircraft without permission from the pilot. After receiving permission from the pilot, the responder may approach the aircraft, but should always approach from within the pilot's view and always from the downhill side if on a grade. Figure 7–5 depicts safe and unsafe approach areas to helicopters during ground operations. Should a rescue basket be used, static electricity is a safety concern. The basket should touch the ground before a rescuer on the ground touches it.

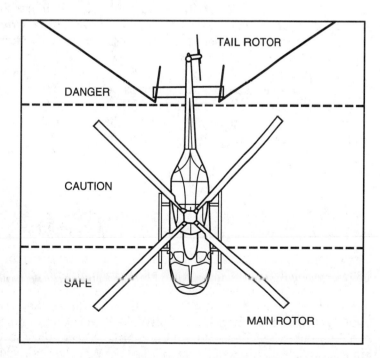

Figure 7–5 *Approach helicopters only from the safe zone.*

CIVIL DISTURBANCES

There have been a number of **civil disturbances** in the United States over the years. Emergency service responders are often on the front lines during these events, either for the treatment of injured or for fire suppression. These incidents require a great deal of coordination and cooperation with law enforcement.

Emergency response personnel operating during civil unrest are exposed to many hazards, including gunfire, being assaulted with objects that have been thrown, and increased hazards at fires due to the use of accelerant. Personnel may minimize some of these risks through prioritization of incidents, police protection, and additional PPE.

Incidents should be prioritized according to the potential risks involved. If a life threat is not present, the risk to personnel should be closely examined before the responders become the life safety concern. If the incident is a priority incident and units must respond, the police should either escort the units or ensure the area is clear prior to fire or EMS response. Emergency response units may be put into task forces in order to respond in higher numbers. This arrangement might be two engines, a ladder, a medic unit, and a battalion chief, with police escort. In the case of priority building fires, units should not commit to long operations, but instead hit fires with booster tanks and master streams. This tactic allows for a fast retreat, should the situation dictate.

■ **Note**
Police protection is a critical component of the operational plan at civil disturbances. This protection should be planned before an event.

Police protection is a critical component of the operational plan at civil disturbances. This protection should be planned before an event. The police often have intelligence as to where problem areas may be and what might be expected. This information should be communicated to the fire and EMS agencies. Communications and a unified command system will ensure that responders are working under the same plan and all agencies have the same information.

Some agencies that do not normally wear body armor may elect to issue the body armor for use during civil unrest. Although SCBA would provide the necessary protection, emergency responders may wish to wear masks to limit their exposure to tear gas. These masks are lighter than SCBA and provide protection from the irritation of tear gas.

TERRORISM EVENTS

The events in New York and Virginia that shocked the nation on September 11, 2001, emphasized that firefighters, EMS responders, and law enforcement personnel are the nation's first responders for terrorism events. Emergency responders were aware of this duty long before these events as they responded during the 1990s to explosions that shook the World Trade Center and the Federal Building in Oklahoma City. In an Atlanta bombing, a secondary device detonated after emergency responders were at the scene. Whether the secondary device was intended for responders is unknown, but the potential exists that it was. No other type response requires a more coordinated effort between agencies. As described in Chapter 4, preevent interagency planning is a necessity, including agreements on communication and the incident management (command) structure.

Terrorism can occur anywhere at any time and an organization's health and safety program must include considerations for health and safety in responding to these events. Terrorism events are extremely complex, therefore a complete study of safety at such events is well beyond the scope of this book; however, some general safety considerations and information is provided.

Terrorism acts, using weapons of mass destruction (WMD), can be classified into five general categories. Generally, terrorism acts are chemical, biological, radiological, nuclear, or explosive (CBRNE) in nature. Emergency responders must, at a minimum, be provided with awareness training for terrorism events. Unlike a fire or vehicle crash in which the problem is evident on the arrival of responders, WMD events may not present so obviously. Biological and radiological events require measurement of air quality very early in the incident, therefore the responders should be equipped with meters to determine air quality. Many WMD events will require the same safety precautions that one would take when responding to a hazardous materials incident. Therefore, proper PPE must be available to all responders including appropriate respiratory protection. Responders should take a cautious approach as described earlier in this chapter, watching for clues that could indicate a terrorism event. Text Box 7–5 lists some of these clues.

Text Box 7–5   *Clues that the event might be WMD related.* (Courtesy FEMA Emergency Response to Terrorism Job Aid.)

- Is the response to a target hazard as identified in the risk analysis?
- Has there been a threat?
- Multiple victims—all with same signs and symptoms or injuries?
- Are responders victims?
- Are hazardous substances involved?
- Has there been an explosion?
- Has there been a secondary attack/explosion?
- Does review of dispatch information provide clues?
- Any physical signs present?
  - Debris field
  - Mass casualty/fatality with minimal or no trauma
  - Severe structural damage without an obvious cause
  - Dead animals and vegetation
  - System(s) disruptions (utilities, transportation, clouds)

Responders should be trained and prepared for mass decontamination activities. Responders must also have antidote kits available that are for use with some nerve agents.

Responders must remember that the terrorism act may be designed to create a media event; emergency responders can easily be targets. The are a number of resources available to the safety and health program manger for training in safe operation at these terrorism events, including training from the FEMA/United State Fire Administration, the Department of Justice, and numerous textbooks. As first responders, the emergency service will be the first agency on scene of a terrorism event and we must ensure our safety as well as that of the public.

NATURAL DISASTERS

Emergency responders are the first responders to natural disasters, including earthquakes, hurricanes, tornadoes, floods, wildfires, severe thunderstorms, and flooding. Because emergency service responders will be going to these events, they must also be given consideration in the health and safety plan. Risk assessment of natural disasters is essential for the local department. Because certain areas of the country are more prone to different types of disasters, it is necessary to develop the safety and health program around the threat. It is unlikely that a firefighter in Ohio would have to be prepared for a hurricane, but should be very prepared for tornadoes. A Florida firefighter has a good chance of responding to the aftermath of a hurricane, but less chance of responding to an earthquake.

In some ways, like the terrorism events described in this chapter, response to natural disasters can be very complex, although many of the hazards faced will be similar regardless of the type. Table 7–1 provides an overview of the hazards associated with natural disaster, the risks, and safety recommendations.

After a natural disaster, emergency responders will be working with numerous other agencies during the response and recovery phase. For the safest possible working environment, interagency coordination with a number of agencies. including emergency management, law enforcement, public works, local utility companies, and in many cases the military or national guard, is important. Preevent relationships should be established so that the responders understand each others' roles, responsibilities, and limitations.

Natural disaster operations are long-term and often widespread events. Concepts presented elsewhere in this textbook—incident management, accountability, and rehabilitation—must have priority in the operation.

Table 7–1 *Hazards associated with natural disaster, their risks and safety recommendations. Adapted from NIOSH Publication at http://www.cdc.gov/niosh/emhaz2.html#haz1.*

Hazard	Risks	Safety Recommendations
Massive piles of construction and other types of debris, unstable work surfaces	Traumatic injuries, including serious fall injuries, from slips, trips, and falls or collapsing materials	1. Ensure that surfaces are as stable as possible. 2. Use alternative methods such as ladders to access unstable work surfaces. 3. Ensure that responders have a full array of PPE, including safety shoes with slip-resistant soles. 4. Ensure that responders use fall protection equipment with lifelines tied off to suitable anchorage points whenever possible.
Excessive noise	Communication and temporary hearing loss	Use hearing protection devices
Breathing dust containing asbestos (from pulverized insulation and fire-proofing materials) and silica (from pulverized concrete), which are toxic	Short term: irritation of eye, nose, throat, and lungs Long term: Chronic effects may depend on the extent and the duration of exposure	1. Responders should be protected from breathing dust. 2. Respiratory protection: An N-95 or greater respiratory protection is acceptable for most activities.
Heat stress from wearing encapsulating/insulating bunker gear or doing heavy work in a hot, humid climate	Significant fluid loss that frequently progresses to clinical dehydration, raised core body temperature, impaired judgment, disorientation, fatigue, and heat stroke	1. Adjust work schedules, rotate personnel, add additional personnel to work teams. 2. Replenish fluids (1 cup water/sports drink every 20 minutes) and food (small, frequent, carbohydrate meals). 3. Monitor heart rate. If over 180 beats per minute minus age for more than a few minutes, stop work and rest immediately. 4. Provide frequent medical evaluation for symptoms and signs of heat stress, such as altered vital signs, confusion, profuse sweating, and excessive fatigue. 5. Provide shelter in shaded areas and the ability to unbutton and remove bunker gear.
Confined spaces (limited openings for entry and exit, unfavorable natural ventilation)	Low oxygen, toxic air contaminants, explosions, entrapment, death by strangulation, constriction, or crushing	1. Purge, flush, or ventilate the space. 2. Monitor the space for hazardous conditions. 3. Lock out/tag out procedures for power equipment in or around the space. 4. Use appropriate PPE such as a SCBA 5. Light the area as much as possible. 6. Establish barriers to external traffic such as vehicles, pedestrians, or other hazards. 7. Use ladders or similar equipment for safe entry and exit in the space. 8. Use good communications equipment and alarm systems. 9. Have rescue equipment nearby.

(continued)

Table 7–1   *(Continued)*

Hazard	Risks	Safety Recommendations
Potential chemical exposures	Eye, nose, throat, upper respiratory tract, and skin irritation; central nervous system depression, fatigue, loss of coordination, memory difficulties, sleeplessness, mental confusion	1. Fire Fighting: Use SCBA with full face piece in pressure demand or other positive pressure mode. 2. Entry into unknown concentration: Use SCBA gear. 3. Rescue operations with fumes present: Use gas mask with front-mounted organic vapor canister (OVC) or any chemical cartridge respirator with an organic vapor cartridge.
Traumatic stress after horrific events	See Chapter 8 for information on critical incident stress	See Chapter 8 for information on critical incident stress management.
Electrical, overhead power lines, downed electrical wires, cables	Electrocution	1. Use appropriately grounded low-voltage equipment. 2. Interact with power company to ensure downed lines are not energized.
Carbon monoxide risk from gasoline- or propane-powered generators or heavy machinery	Headache, dizziness, drowsiness, or nausea progressing to vomiting, loss of consciousness, and collapse, coma, or death under prolonged or high exposures	1. Use CO warning sensors when using or working around combustion sources. 2. Shut off engine immediately if symptoms of exposure appear.
Eye injuries from dust, flying debris, blood	Bloodborne pathogen infection, eye injury	1. Use goggles or face shield and mask for those handling human remains, recovering deceased. Make sure to cover the nose and mouth to protect the skin of the face and the mucous membranes. 2. Use safety glasses with side shields as a minimum by all responders. An eye wear retainer strap is suggested. 3. Consider safety goggles for protection from fine dust particles, or for use over regular prescription eyeglasses. 4. Any worker using a welding torch for cutting needs special eye wear for protection from welding light, which can cause severe burns to the eyes and surrounding tissue. 5. Only use protective eyewear that has an ANSI Z87 mark on the lenses or frames.
Flying debris, particles, handling a variety of sharp, jagged materials	Traumatic injuries ranging from minor injuries requiring first aid to serious, even disabling or fatal traumatic injury	1. Use safety glasses with side shields as a minimum. 2. Consider safety goggles for protection from fine dust particles. 3. Educate responders regarding safe work procedures before beginning work. 4. Provide responders with a full array of PPE, including hard hats, safety shoes, eyeglasses, and work gloves. 5. Ensure that responders do not walk under or through areas where cranes and other heavy equipment are being used to lift objects.

Table 7–1   (*Continued*)

Hazard	Risks	Safety Recommendations
Work with numerous types of heavy equipment, including cranes, bucket trucks, skid-steer loaders, etc.	Traumatic injury, including serious and fatal injuries, due to failure or improper use of equipment, or responders being struck by moving equipment	1. Train responders to operate equipment correctly and safely. 2. Ensure operators are aware of the activities around them to protect responders on foot from being struck by moving equipment. 3. Ensure that responders do not walk under or through areas where cranes and other heavy equipment are being used to lift objects.
Rescuing victims, recovering deceased, handling human remains, contact with surfaces contaminated with blood and body fluids	Blood, bloody fluids, body fluids, and tissue are potential sources of bloodborne infections from pathogens including Hepatitis B, Hepatitis C, and HIV.	Standard precautions (universal precautions) should be strictly observed regardless of time since death.

SUMMARY

For the purpose of this section, response to specialized incidents includes responses to hazardous materials incidents, technical rescues, helicopter operations, civil disturbances, terrorism, and natural disasters. Each of these requires specialized response and produces numerous unique safety concerns for the responders. While some of these types of incidents may not be handled by every organization, it would be hard to imagine any organization that does not have the potential to respond to these specialized incident emergencies. In some cases the organization may function as the first responders; in others the response may be with a specialized team. Hazardous materials operations are governed by federal regulations that require responders to be trained to a minimum level, to operate with proper PPE and in a buddy system, and to have site-specific safety plans and an IMS. Training is based on five levels of response. The safety program manager must access the responder's level of training and ensure it is commensurate with the type of operations being performed.

Technical rescue is a specialized area of response that requires specialized training and equipment. Technical rescue includes rescue from structural collapse, trench and excavation rescue, rope rescue, confined space rescue, rescue from vehicles and machinery, wilderness rescue, and rescue from water. Each of these types of technical rescues are disciplines within themselves. From a safety standpoint, although some generalization regarding safety can be made, these require a great deal of expertise on behalf of the responder, the incident commander, and the assigned incident safety officer.

Many departments are using helicopters for fire suppression and, more likely, medivac operations. At a minimum, emergency responders may be required to provide ground support for these operations. Critical ground support safety issues relate to the selection and protection of a landing zone, crowd control, and the approach to the aircraft. Local conditions and the operator of the helicopter dictate specifics regarding these issues.

Response and operations in areas of civil disturbances require close coordination with law enforcement including unified command in the IMS. Through the use of prioritization of calls in the affected area, the use of alternative tactics, task force response, and increased PPE, a level of safety can be provided.

Terrorism events also require the close coordination between law enforcement and the emergency responders. Acts of terrorism may include chemical, biological, radiological, nuclear, or explosive weapons of mass destruction. Intended to create media attention, such incidents often involve mass casualties and may have secondary devices designed to injure emergency responders.

Finally, natural disasters include weather-related events, floods, wildfires, and earthquakes. Although some specific areas of the country are more likely than others to experience specific natural disaster threats, there are general hazards and risk that must be considered in the safety and health program.

Concluding Thought: Emergency responders are required to handle much more than fire and EMS incidents. The issues related to safety increase with the various other incidents to which we must respond.

REVIEW QUESTIONS

1. A person who normally responds to releases of hazardous substances and takes offensive action by plugging a leak should be trained to what level?

   A. First responder operations

   B. First responder awareness

   C. Hazardous materials technician

   D. Incident commander

2. A person who normally responds to releases of hazardous substances and takes no offensive action other than notification to start the response plan should be trained to what level?

   A. First responder operations

   B. First responder awareness

   C. Hazardous materials technician

   D. Incident commander

3. What does the acronym RAID stand for?

4. List and explain three pieces of information available to help the responder identify a released hazardous substance.

5. In terms of terrorism, what does the acronym CBRNE stand for?

6. What seven areas of rescue are covered by the term *technical rescue*?

7. In general, what is the minimum space requirement for a helicopter landing zone?

8. What action must be taken before approaching a helicopter with a rescue basket deployment?

9. What PPE should be worn during helicopter landing zone operations?

10. List three hazards present during a civil disturbance.

ACTIVITIES

1. Review your department's SOPs regarding the response to hazardous materials incidents and compare the SOP to what really happens in the field. Do the responders comply? Are they trained to the level required for the response?

2. Access your department's response levels to technical rescue incidents. What services or types of responses are provided? What safety issues have been identified and what procedures are in place to reduce the hazards?

3. Access the potential for civil disorders and terrorism events in your community. What procedures are in place to ensure a safe response?

4. Review the types of natural disasters that are likely to occur in your response area. Do the procedures in place consider responder safety and health?

Chapter

8

POSTINCIDENT SAFETY MANAGEMENT

Learning Objectives

Upon completion of this chapter, you should be able to:

- List the safety and health considerations when terminating an incident.
- Describe the demobilization process.
- Compare the concept of first in/last out with first in/first out.
- Explain the need and the process used for postincident analysis.
- List the components of a postincident analysis.
- Describe the advantages of a critical incident stress management program.
- List the key components in a critical incident stress management program.

CASE REVIEW

On Saturday, March 13, 2004, shortly after 9:00 A.M. the City of Pittsburgh Fire Department responded to a structure fire at the Ebenezer Baptist Church on Wylie Avenue in the Hill District. When the first-due units arrived, they reported heavy smoke inside the large church and believed that the fire was in the basement. Crews began an offensive attack by advancing attack lines to the basement, where they were met by heavy smoke and heat conditions. As they continued to search for the seat of the fire, a second alarm was transmitted for additional resources.

Shortly thereafter, a third alarm was requested and crews were ordered out of the building as conditions continued to deteriorate. As the crews were leaving the building, an apparent backdraft occurred injuring several firefighters. Defense operations continued as a fourth alarm was requested. A number of master streams and handlines were deployed around the building for the defensive firefight.

Approximately three hours later, the bulk of the fire was knocked down and crews began to place handlines in service to extinguish hot spots. Then, without warning, the bell tower collapsed, killing two firefighters and injuring approximately twenty-nine.

Although this tragic event occurred while units were still at the scene, it emphasizes the point that injuries and fatalities can occur as the incident is brought under control. Postincident operations are an important part of the overall safety and health program.

INTRODUCTION

The last step in the safety and health program consideration in terms of incident response is the consideration for postincident activities. Included in this chapter are suggestions for terminating the incident, postincident analysis, and critical incident stress management. Postincident safety and health considerations can be easily overlooked if not integrated into the total safety and health program. In other words, responders tend to let their guard down after the incident is over. Therefore, the safety officer and supervisors must be trained, and safety during this phase should be emphasized and committed to as part of the program.

■ **Note**
Responders tend to let their guard down after the incident is over.

INCIDENT TERMINATION

Incident termination can be comprised of three stages: demobilization, returning to the station, and postincident analysis. Postincident analysis is discussed in detail in the next section of this chapter.

Demobilization

Demobilization is the stage in the incident when the incident commander in a large incident, with the recommendation of the planning section, evaluates the

demobilization
the process of returning personnel, equipment, and apparatus after an emergency has been terminated

first in/last out
the common approach often used at an emergency scene; basically, the first arriving crews are generally the last to leave the scene

■ Note
Lately some departments have begun to change this approach, using first in/first out as a demobilization tool.

standardized apparatus
apparatus that has exactly the same operation and layout of other similar apparatus in a department, for example, all of the department's pumpers would be laid out the same, operate the same, and have the same equipment, useful for the situations when crews must use another crew's apparatus

on-scene resources and compares them with the current situation. Using a fire scenario, demobilization usually occurs when the fire has been brought under control and some of the firefighter resources can be placed available and returned to stations. Demobilization, particularly at very large incidents, requires planning and coordination. The National Interagency Incident Management System provides a form, Form ICS-221, to assist in the coordination of demobilization (see Figure 8–1).

There are numerous safety and health issues that the incident commander should address during this stage and that should be part of the department's operating procedures and safety policies.

One consideration is the sequence in which operating crews are released. The fire service has historically used the **first-in/last-out** approach. Very often, the unit that is first on the scene is the last unit to leave. Lately some departments have begun to change this approach, using first in/first out as a demobilization tool. This process is good because it permits the units that arrived first, and probably have worked the longest and generally under the worst conditions, to be relieved first. Some arguments are made regarding this process as the first arriving apparatus are normally committed to firefighting and have ladders raised to the building or hose connected. This argument can be overcome when departments have **standardized apparatus** and equipment. In this case it should make no difference that a first-arriving crew could later leave in another crew's apparatus. The first-in/

Figure 8–1
Demobilization is the stage in which units and personnel begin to leave the scene.

first-out approach is a very important consideration during periods of extreme heat or cold and high humidity.

The incident commander and the planning officer should also consult with the on-scene safety officer during the demobilization process. The safety officer may be more aware of crews that need to be relieved or that have operated for extended periods. The safety officer can obtain some information from the rehabilitation officer who should have been monitoring the status of members of each crew.

The use of rapid intervention crews or teams (RIC) was discussed in Chapter 5. During demobilization, the incident commander should make an informed decision as to when the RIC is no longer needed. It is not uncommon for injuries to occur long after the initial fire fight is over. In fact, some disastrous collapses in which crews have been trapped and killed have occurred during the overhaul phase of a fire. One such incident is noted in the case review of this chapter.

■ Note
The supervisor or senior person assigned to the crew must ensure the readiness of the team for the next incident.

Once released from the emergency scene, or in the case of a medical incident once the call has ended, there are safety concerns as well. The supervisor or senior person assigned to the crew must ensure the readiness of the team for the next incident. After any incident, the crew members should be questioned about injuries they may have incurred or any feelings about the incident that may prompt the supervisor to start the critical incident stress management process. In the case of an incident in which crew members have the potential to be exposed to infectious diseases, the crew should do as much personal decontamination as possible at the scene or at the hospital prior to returning to the vehicle. At this point the supervisor or company officer must make a decision as to whether the unit can be placed in an available status or if crew members or equipment are not ready. It is not uncommon, in the emergency service field, to check equipment and stock after a call to ensure readiness. The human safety elements cannot be overlooked in this process.

Returning to Station

critical incident stress
stress associated with critical incidents, such as the injury or death of a coworker or a child

The supervisor should also be alert to signs of **critical incident stress.** Vehicle operators must maintain an alertness to the road (see Figure 8–2). Reviewing the NFPA injury and death statistics, the category for vehicle crashes is titled "responding/returning." Injuries do occur during the returning-to-station stage of the incident. Returning to the station often provides the first chance for the crew to discuss the incident among themselves.

Once back at the station, the apparatus is made ready for the next emergency, supplies are replaced, and equipment cleaned. Again, the safety program must incorporate the human needs at this point. The supervisor should be alert for signs of fatigue and the crews should be assessed again for any injuries or illness that may be attributed to the incident. Remember, emergency service responders often do not complain because they see minor injuries as part of the job. The supervisor should be alert to visual signs such as limping or favoring an arm or leg, cuts,

Figure 8–2 *Safety is also a consideration during the trip back to the station and putting equipment back on the apparatus.*

postincident analysis
a critical review of the incident after it occurs; the post-incident analysis should focus on improving operational effectiveness and safety

bruises, and general differences in the responder. Nine out of ten times the responders are just tired, but sometimes an injury can be uncovered and reported simply by asking the right question.

Once the equipment and crews have been taken care of, the unit is readied for the next assignment. Before long, however, the supervisor should sit down and discuss the incident with the crew in an informal manner. This is the first step in the **postincident analysis**.

POSTINCIDENT ANALYSIS

A postincident analysis should be done on any incident to which multiple units have responded and operated and any incident involving a serious injury or death. These incidents may include fires, technical rescues, hazmat incidents, multiple casualty incidents, vehicle accidents, and victim entrapments. Often called a critical review, or critique, the postincident analysis (PIA) is a step-by-step look at what happened at the incident, what went right and what went wrong. The components that the PIA should focus on are:

- Resources
- Procedures
- Equipment
- Improving operational effectiveness

The term *critique* is sometimes associated with criticize, therefore the term *postincident analysis* or PIA is used in this text.

There are two steps of the PIA. The first is the informal discussion among the crew after the incident (see Figure 8–3). Although informal, this is a good time to review what each member saw and did and what the unit did as a whole. The supervisor during this discussion should make notes for use at the formal PIA. This process can, and should, also occur in volunteer organizations.

Usually within some fixed time after the incident, all crews that were present at the incident should get together for the formal PIA. This PIA must be attended by all those who carried out a specific task at or for the incident, including those members who filled positions within the incident management system (see Figure 8–4). Often the incident commander will call for and run the PIA. It is better in most cases to have an objective person other than the incident commander facilitate the PIA as this then permits a free exchange of information and allows the

Figure 8–3 *The informal PIA may take place at the kitchen table after the incident.*

Figure 8–4 *The formal PIA should involve all members that operated at the incident.*

incident commander to be more involved in the PIA as opposed to having to lead the analysis. The PIA events should be recorded for future changes in procedure or policies or to reaffirm current ones. Some departments videotape the PIA and then distribute the tape departmentwide so that every member can learn from the incident. At minimum there should be a written form that is filled out and filed for each PIA conducted.

A good procedure for the PIA is to have all the responders discuss, usually using a graphic or plot plan of the incident, what they did on arrival, what worked, and what did not work. If injuries or deaths occurred as a result of the incident, surrounding events should also be noted and investigated. Someone present at the PIA should have a copy of the SOPs available and compare the actions taken to those required by the procedures. For example, if the SOP says that the first-arriving ambulance will set up a treatment area, and in the case in question, the first-arriving ambulance entered the building to fight the fire, then either the SOP needs to be reevaluated or the crew should state the reasons for their actions.

The objective of the PIA should be to highlight positive and negative outcomes of an incident. The PIA can also provide a tool for future training and show a need for changes in procedures and policies or even a need for a new piece of equipment. The PIA should be a positive process, allowing all attendees to have a chance to speak and discuss concerns, particularly those related to short- and long-term safety and health issues. The PIA can also aid in the investigation process into any incident injuries and fatalities and become cause to develop procedures that may prevent them in the future. See Figure 8–5 for the relationship between SOPs, training, operations, and the PIA.

CRITICAL INCIDENT STRESS MANAGEMENT

Stress is a normal part of everyday life; however, too high a stress level leads to negative reactions, such as actual physical illness, job burnout, and lack of productivity. In terms of safety and health, a stress management program is necessary,

stress

the body's reaction to an event; not all stress is bad; in fact, some level of stress is necessary to get a person to perform, for example, the stress associated with a report that is due is often the motivating factor in doing it

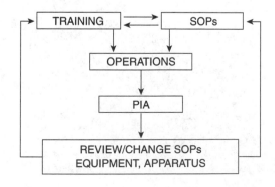

Figure 8–5 *The relationship between the PIA, SOPs, training, and operations.*

including provisions for employee assistance programs as discussed in Chapter 4. This section of the text focuses on after-incident critical incident stress management and the associated health and safety considerations.

Emergency responders are expected to tolerate certain levels of stress. However, some events that are usually high in stress levels are those that are powerful in terms of emotion or a combination of many smaller events. These events are called *critical incidents*. Many years ago, responders were expected to handle these events on their own and were often criticized if they displayed weakness or emotion toward these critical events. Fortunately, this is no longer the case and critical incident stress management has become not only accepted in the emergency service arena, but is a part of most organizations.

Events that typically result in critical incident stress include the following:

- Traumatic death or severe injury to a coworker, particularly those suffered in the line of duty (see Figure 8–6)
- Suicide of a coworker
- Traumatic death or serious injury to children
- Mass casualty events
- Prolonged events
- Death or injury to a person caused by the emergency responder, for example a traffic accident while responding to an incident
- Events in which there is a great deal of media attention
- Personally significant events, for example the death of an elderly man when the responder just lost an elderly family member

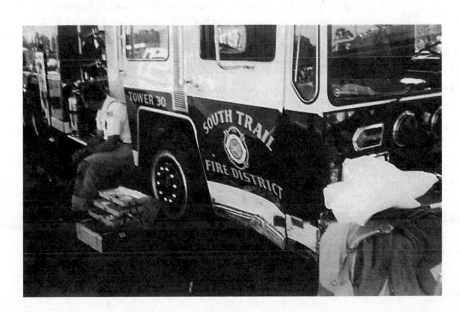

Figure 8–6
Accidents involving coworkers are particularly stressful. CISM should always follow these incidents.

When a critical incident occurs, it is important that a process be in place for both stress defusing and debriefing. This process should include access to the Critical Incident Stress Management (CISM) system. In some cases, this debriefing should take place before the responders have to ask for it, as the responders may not feel they need it or may not ask for fear of showing signs of weakness. Each organization sets procedures for when and how to access the system. Some systems automatically set the process in motion when any of the critical incidents listed takes place or at the request of an incident commander or company supervisor for any event.

Without CISM, the responder may be more likely to develop negative physical, behavioral, and psychological reactions. Unwanted physical reactions include fatigue, sleeplessness, changes in eating habits, and body aches. Behavioral changes may also occur, including changes in activity levels, difficulty in concentration, nightmares, flashbacks, memory problems, and isolation. Psychological symptoms include fear, guilt, sensitivity, depression, and anger. In some extreme cases, if not properly managed, a person suffering from critical incident stress reactions may begin to abuse alcohol and drugs and may resign from employment in the emergency service field. The safety and health program that includes CISM helps the responder deal with the issues associated with the event and hopefully avoid many of the unwanted reactions.

There are different types of CISM both at the scene and after the event. The first should occur at the scene and is termed **peer defusing.** Peer defusing involves an informal discussion about the event by a group of peers who experience the event together. This phase often occurs normally as the incident deescalates and responders get together and discuss the event. The second is the formal debriefing that usually occurs sometime after the incident, maybe even a few days, and is the formal process in which only the responders to the incident get together with a peer debriefer and discuss their roles at the incident and the overall response. Two issues are very important in this phase: (1) All those involved should try to attend, and (2) the peer counselor should have had training in this type of operation.

Very often CISM teams are regionalized, which allows for a peer debriefer from the same occupation but an outside agency to serve in that capacity. It would be difficult for a department to have an inside team, as each debriefing would involve a peer from the same department who was not at the incident. This approach is not good. Good CISM team characteristics include regionalization, a mental health professional as a team member, nonpartisan, and having received appropriate training in CISM. Total confidentiality and relative anonymity are vital elements of a debriefing and the peer debriefer must follow and buy into these concepts, as should the management of the response organizations.

A third type of CISM is on-site defusing, which can be useful at large, long duration events. As they leave the operational area, responders are evaluated by a member of the local CISM team for signs and symptoms of distress and treated appropriately. The fourth type of CISM is demobilization. Demobilization allows

peer defusing
the concept of using a trained person from the same discipline to talk to a emergency responder after a critical incident occurs as a means to allow the responder to talk about his or her feelings about the event in a nonthreatening environment

■ **Note**
All those involved should try to attend, and the peer counselor should have had training in this type of operation.

■ **Note**
Total confidentiality and relative anonymity are vital elements of a debriefing and the peer debriefer must follow and buy into these concepts, as should the management of the response organizations.

responders after operating at a large-scale event to have a buffer period of 30 to 40 minutes before leaving the location. This period can provide a time for nourishment and brief education on the signs and symptoms of critical incident stress.

CISM is a relatively new concept in emergency response. Although the incidents have occurred forever, it was not until the 1980s that incident stress was recognized as a problem. The safety and health of the department members and their longevity in the response system depend on CISM. It is a necessary component to any safety and health program.

SUMMARY

Safety and health after the incident are sometimes overlooked. During this phase of the response, there is potential for injuries to occur and, therefore, this must be considered in the safety and health SOPs. Postincident safety considerations begin with the termination of the incident, at which time the incident commander makes a decision about the demobilization of resources, and these resources return to their stations. On a single unit medical-type incident, the incident is terminated once the patient has been handed over to the emergency department or to a transport provider. It is important during this phase of the operation that the same consideration be given to the human components of the crew that is given to the equipment and supplies.

Part of the postincident safety considerations is postincident analysis (PIA) and critical incident stress management (CISM). The PIA can begin when the unit returns to the station and the superiors have an informal discussion with the crew regarding the incident. Later, a formal PIA must be

conducted using a prescribed format and documented for future use in training or for updating SOPS. All personnel who responded to the particular incident must be included in the formal PIA.

Another safety and health concern postincident is critical incident stress management. Critical incident stress management programs are integral to the safety and health program. Teams should be available to assist members following incidents that generate high levels of stress. Often these teams are formed on a regional basis to allow for peer reviewers from outside of one's own agency. Supervisors must have training in and be aware of common incidents that can lead to critical incident stress reactions. All personnel must be aware of available CISM programs and how to access them at any time, given the need.

Concluding Thought: Many essential safety and health program functions occur after the incident itself, but are just as important as those that occur before or during an incident.

REVIEW QUESTIONS

1. What are the three stages in terminating the incident?

2. Differentiate between the first-in/first-out and the first-in/last-out approaches. List advantages and disadvantages of each.

3. It is best to allow the incident commander to facilitate a PIA.

   A. True

   B. False

4. Using a supervisor from within the organiza-

tion as a peer debriefer is an important step in the CISM process.

   A. True

   B. False

5. List three qualities of a good CISM team.

6. The RIC can be released once a fire is declared under control.

   A. True

   B. False

7. Describe the relationship between training, SOPS, operations, and the PIA.

8. Inasmuch as possible, decontamination should occur before leaving the scene.

   A. True

   B. False

9. Who should be present at a formal PIA?

10. The informal PIA should occur on return to the station before the unit is placed back in service.

    A. True

    B. False

ACTIVITIES

1. Review your department's SOP for demobilization at emergency incidents. Which process is used, first in/first out or first in/last out? Consider changes. Is the approach used in your department working? What roadblocks might prevent a first-in/first-out procedure?

2. Develop a postincident analysis SOP for your department. If there already is one, how effective is it? Are the objectives noted in this text being achieved?

3. Research information on CISM teams in your area. Have personnel had training in recognizing critical incident stress reactions and how to access the team?

Chapter 9

PERSONNEL ROLES AND RESPONSIBILITIES

Learning Objectives

Upon completion of this chapter, you should be able to:

- List the roles and responsibilities of the individual responders and their relationship to the overall safety and health program.
- List the roles and responsibilities of the supervisors and their relationship to the overall safety and health program.
- List the roles and responsibilities of the emergency service management and their relationship to the overall safety and health program.
- List the roles and responsibilities of the incident commander and his or her relationship to the overall safety and health program.
- List the roles and responsibilities of the safety program manager and his or her relationship to the overall safety and health program.
- List the roles and responsibilities of incident safety officers and their relationship to the overall safety and health program.
- List the roles and responsibilities of the safety committee and its relationship to the overall safety and health program.

CASE REVIEW

The department had everything in place. The board of fire commissioners had decided several years earlier that the department had grown to a size that the position of training/safety officer was needed. The department did a promotional process and the assignment was made.

The training officer immediately set out to improve the department's training and safety program. In terms of a safety program, the training/safety officer started a safety committee, performed a safety audit on the department based on the *NFPA 1500* standard, and introduced the concept of risk management. After laying this basic ground work, the officer, with the aid of the safety committee, developed a number of very good programs.

The training/safety officer made a mistake during the implementation of the programs. Specifically, the training/safety officer failed to identify the roles of the various employees in the department. Some of the feedback he was getting was that, although the statistics for the department were getting better, many of the operations people on a particular shift were not following some of the basic safety procedures and policies such as seat belt use and using PPE on medical

alarms. The safety officer was concerned that if the procedures were not enforced on some responders, other members would follow the procedures only when convenient.

But who was at fault? The safety officer began at the first-line supervisor level and through training and the use of case studies reemphasized the importance of the program and the importance of enforcement. The training sessions were well received in most cases. After seeking support from the department management, the safety officer developed a written safety statement signed by the fire chief that made it clear that safety and health in the operations of the department was a top priority. This policy helped the company officers and individual responders in understanding the importance of the procedures and preventive action that were being required. Further, the policy emphasized that noncompliance could result in discipline. Over time, the department's attitude toward safety changed for the better. The importance of understanding the various roles and responsibilities of the personnel was clear to the newly hired training and safety officer. Everyone is a stakeholder in personnel safety and must understand his or her role.

INTRODUCTION

As part of putting a safety program together, the safety program manager must understand the various roles that the members of the organization play. Who exactly is responsible for the safety and heath program within an organization? After reviewing the information that has been presented thus far, the answer should be clear—everyone. Safety and health responsibility extends from the new individual member to the top of the organization, whether that top is the chief or other titled director. Recalling the SAFEOPS approach from Chapter 1 (see Table 9–1), it can quickly be seen that many different persons in the organization have responsibility for one or more components of SAFEOPS. The interplay of the various groups and individuals in the safety and health program can be compared to a spoked wheel, and the wheel is only as strong as its weakest spoke (see Figure 9–1).

Table 9–1 *Primary responsibility for the components of the SAFEOPS approach.*

Component	Primary Responsibility
Supervision	Incident commanders Supervisors Safety program manager Management Incident safety officers
Attitude	Individual responders Supervisors Management Incident commanders Safety program manager Incident safety officers Safety committee
Fitness	Individual responders Supervisors Management Incident commanders Safety program manager Incident safety officers Safety committee
Education/Wellness	Individual responders Supervisors Safety program manager Management
Organizational involvement	Safety committee Safety program manager Management
Procedures	Individual responders Incident safety officers Incident commanders Safety committee Safety program manager Management
Standards/Regulations (compliance with)	Incident safety officers Incident commanders Safety committee Safety program manager Management

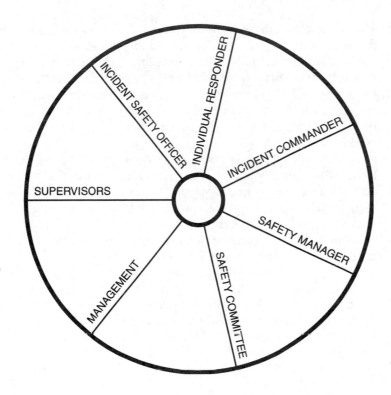

Figure 9–1 *The interrelationship of roles can be likened to a spoked wheel. The wheel will only be as strong as the weakest spoke.*

Although there are differing responsibilities in the various organizational levels, the goal of the program remains the same. This chapter examines these various levels and the associated roles that the persons in them may play.

INDIVIDUAL RESPONDERS

Individual responders are probably the most important link in the safety and health program. This may sound unusual, however, the individual responders can have a great impact on the entire program. It is not important how many policies are written, how many safety officers are assigned, or how much money is put into a program, without compliance and a safety attitude on behalf of the responders, the program will not be effective.

The "A" in SAFEOPS represents attitude. Attitude may well be the most important obstacle to overcome when dealing with safety and the individual responder. Although changing, the historical belief that injuries and death go with the occupation is still part of many responders' attitudes. The belief or attitude that it will not happen to me is also human nature. There was once a line in

■ Note
It is not important how many policies are written, how many safety officers are assigned, or how much money is put into a program, without compliance and a safety attitude on behalf of the responders, the program will not be effective.

a war movie in which the actor was shot. The actor's line was "I'm shot. I never thought I would get shot!" One would think that a person fighting a battle in a war could assume that the potential for being shot was reasonably high, yet he was surprised that it really happened to him. We see this same belief with people, whether involving crime, car accidents, or, in some cases, being injured at work. In order for a responder to want to operate safely and comply with procedures and polices, the responders must realize that the potential for injury to him or her is real and could happen.

Individual responders must also realize that they are members of a team. Depending on the type of response unit, this may be a two-person team on an ambulance or a five- or six-person team on a fire unit. Although an individual, the responder's action almost always has some impact on the other members of the team. For example, if a firefighter does not wear all of the necessary protective equipment and is overcome with smoke, the rest of the team has to rescue the firefighter, and the entire team is affected.

The individual responder should also be willing to take an active role in the safety and health program. This may involve making suggestions, working on the safety committee, and using good safety practices. It is important that the more senior responders set an example, particularly when training a new recruit. Text Box 9–1 describes some individual responder dos and don'ts.

Text Box 9–1 *Dos and don'ts for individual responders.*

Do Be:

- *An active team player.* Be prepared to fill your role and watch out for team members, including following safety-related procedures. If a team member does not have protective gloves on when starting an IV, remind her.

- *A good communicator.* This applies to both listening and speaking. When working as a team, communicate with other team members. For example, when raising a ladder or lifting a stretcher, maintain contact with team members and ensure that the team makes the lift together on command.

- *A leader.* Display safety leadership. Set the example.

- *Aware of surroundings.* Maintain a constant awareness of surroundings for both yourself and the team. Look for danger signals, communicate concerns.

- *A responder who works within individual abilities.* If you know you cannot do something alone, get help. Countless back injuries associated with lifting could have been avoided when others were available to help.

- *Together.* Stay with your assigned team; no freelancing.

- *A responder with a Safety Attitude!*

Don't Be:

- *Preoccupied.* Emergency responders are called on at a moment's notice. Be thinking about the incident and the potential outcome. Put everything else out of your mind.

- *Complacent.* Do not let the routine nature of a medical incident allow you to let your guard down. Do not assume that the fourth fire alarm today in the same building is not a fire.

- *Surprised.* Many events that occur at emergency incidents are predictable—a disorderly crowd in a bar, a building collapse after exposure to fire. Predict and be prepared.

- *A responder who attempts to do things outside of personal abilities.* If you cannot swim, do not jump in the water to rescue someone.

- *Maintain an attitude that it will not (cannot) happen to me.* Maintain a Safety Attitude.

SUPERVISORS

Although many members of the organization fall into the category of supervision, for the purpose of this chapter, the term supervisor is used to describe first-level or frontline supervision in the emergency response organization. For fire service departments this may be a company officer, lieutenant, or captain assigned to a firefighting unit. In the case of sole emergency medical services (EMS) providers, it may be a senior paramedic or crew chief. Regardless of title, the safety and health roles and responsibilities for the first-line supervisor remain the same (see Figure 9–2).

■ **Note**
The first-line supervisor is the grassroots person who supervises the team.

The first-line supervisor is the grassroots person who supervises the team. The supervisor ensures that team members stay together, maintains accountability, sees that seat belts are fastened prior to apparatus moving, and ensures that personal protective clothing for the job at hand is worn. The supervisor may also be responsible for inspection of protective clothing and seeing that regular maintenance procedures are followed.

■ **Note**
The first-line supervisor is the link between the organization's management and the individual responder.

The first-line supervisor is the link between the organization's management and the individual responder. The supervisor is responsible for seeing that procedures are followed and are understood by team members. The supervisor also communicates safety concerns from the team to higher management levels in the department.

The first-line supervisor generally has risen through the ranks and completed some type of performance-based promotional process. In some cases, supervisors might be assigned based solely on seniority. In either case, the first-line supervisor should have the necessary mix of experience and education to recognize dangers inherent in the occupation, whether on an emergency scene, responding, or in the station. The supervisor must apply this knowledge and take

Figure 9–2 *Fire department company officers have a number of responsibilities associated with the safety of their crews.* (Courtesy Monroeville Volunteer Fire Department Company #4.)

responsibility for the crew in terms of safe operations, including enforcing organization operating procedures, following laws, and maintaining awareness of team location and any dangers present. The supervisor must also follow the practices noted previously for individuals, with specific emphasis on leadership and setting the example.

EMERGENCY SERVICE MANAGEMENT

Somewhere in the organization above the first-line supervisory level is the organization's management. This might be the fire or EMS chief, or the director of emergency services, depending on the type and structure of the organization. For the purpose of this section, emergency service management is used to define the very top levels of the organization, or senior management staff. The safety program manager, regardless of rank or position should be a part of the senior management structure, but is discussed later in this chapter (in the section titled Safety Program Manager).

Senior management also has many roles in the safety and health program. While having the roles and responsibilities as individuals and as supervisors, the senior management staff often controls the financial resources and gives final approval for policy and procedure implementation.

The senior management staff must give the safety program priority in terms of support, both financial and administrative. This support must be realized by the individual responders at lower organizational levels. For example, the paramedic on the street must know that the EMS chief has a commitment to provide a safe work environment with the employees' health as a driving factor. Have you ever heard the statement, "We will do such-and-such because it is required by OSHA"? A better statement might be, "We will do such-and-such, because, after analysis, we have found it to another measure of safety provided to our members." One way of communicating this commitment is through a written organizational safety policy. An example policy is shown in Text Box 9–2.

Text Box 9–2 *Organization safety policy.*

Management Commitment and Involvement Safety Policy

It is the policy of our department to follow the highest safety and health standards. Safety does not merely occur by chance, but occurs as the result of commitment and participation. Our firefighters are our most important assets. Lost time from the job due to accidental injuries is costly to everyone. Our objective is a safety and health program that will eliminate or reduce the number of accidents and injuries to an absolute minimum.

The responsibility for a safe and healthful environment is shared throughout the fire department. The safety officer provides my ongoing direction and guidance in safety matters. I, in conjunction with the department officers, am responsible for training. All officers/supervisors shall execute and enforce these safety rules, policies, and procedures with the utmost regard for the safety and health of our firefighters.

Our departments's policy is that employees make suggestions and report accidents and unsafe conditions to their supervisor and know their rights not to perform unsafe work tasks, without fear of reprisal. I provide the leadership for the safety and health program's effectiveness and improvement to promote safe working conditions through employee feedback, safety committee reviews, and program funding. The safety officer responsible for the safety and health program for our organization is:

Ongoing safety program activities shall include:

- A safety committee with firefighter representation
- Safety orientation, education, and training
- Safety committee review of safety rules, policies, and procedures
- Appropriate corrective actions on accident reports and revising safety rules when appropriate

Firefighters are responsible for following all aspects of the safety and health program and for compliance with all rules and regulations. Responsibility for continuously practicing safety while performing their duties is a condition of employment. All violations of this manual shall be investigated and disciplinary action applied as warranted.

Fire Chief _____ Date _____

INCIDENT COMMANDER

Most injuries in the emergency service occur while operating at various types of incidents. The standards, and in some cases the regulations, require that incidents, regardless of type be managed with an IMS. At the highest level of the IMS is the incident commander (IC). The fact that most injuries occur at the incident scene and that the IC is in charge of the incident scene leads to the conclusion that the IC has a great deal of responsibility in ensuring safety at the incident. The IC is charged with setting the overall incident strategy and strategic goals and assigning tactics to meet these goals (see Text Box 9–3). The IC must balance risks involved in meeting the strategic goals while maintaining a constant regard for team safety. A careful, but quick, risk benefit analysis must be performed before deciding on a strategy.

The IC must first decide on the basic strategy—offensive or defensive—to be employed. The IC should make this determination based on past experience, education, and risk assessment. The IC must rely on information that may or may not be accurate at the time he receives it. Common information is the type of building, potential for occupants, the progression of the fire, and the resources available. For example, a vacant building with moderate fire involvement might immediately lead one to believe that a defensive attack is the strategy of choice. But if this building were in the middle of a block of closely built, occupied buildings, an offensive attack might be ordered.

Incident commanders must have a good knowledge of fire dynamics, building construction, and accepted tactics. They must also have a good level of field experience. Incident commanders can rely on past incidents to provide knowledge as to what is effective and what is ineffective. They can also learn from attending postincident analysis of other incidents and applying the lessons learned to future incidents. Although the decision-making process taught in many officer programs relies on a several-step decision-making process, in actuality, research shows that ICs use what is known as the recognition-primed decision-making (RPD) making process at emergency scenes. Basically the RPD theory says that ICs do not have

Text Box 9–3 *Strategic differences.*

> The incident commander must evaluate the risks of offensive and defense fire attacks.
>
> *Offensive Attack Strategy.* Firefighters are in close contact with the fire and exposed to all the inherent dangers, including burns, falling objects, being lost, being overcome with toxic gases, and building collapse.
>
> *Defensive Attack Strategy.* Firefighters are outside the immediate area of fire danger. Most protective equipment is still required. Dangers still exist, such as building collapse, falling objects, and exposure to heat and fire products.

time at an emergency to go through a several-step process with selection of alternatives, but instead, generally arrive and immediately begin to take control of the incident based on past experience with similar incidents. If something worked ten other times at a bedroom fire, it will work again. For the IC to utilize this process and to provide for sound strategic plans, simulation training is necessary, so the IC can be trained to react to predicable events at the emergency scene.

SAFETY PROGRAM MANAGER

The safety program manager is the focal point for safety-related activities in the organization. As stated previously, the safety program manager should be a member of the senior staff. The program manager has the overall responsibility for overseeing the program and all associated components. *NFPA 1521* requires that, even though the fire chief has the responsibility for the department's safety and health program, a health and safety officer shall be assigned to manage the fire department occupational safety and health program.

The roles and responsibilities of safety program managers are summarized in Text Box 9–4. Depending on the size of the organization, the safety program manager may have a staff of assistants, or, in a smaller organization, may fulfill these roles in addition to other duties, such as being the training officer. In either case, the safety program manager must have sufficient experience and education in risk management principles, cost/benefit evaluation, safety and health issues that face responders, infection control, and emergency service operations to be fully effective. In addition, *NFPA 1521* requires that the safety program manager meet the requirement of Fire Officer Level I in *NFPA 1021*.

Text Box 9–4 *The role and responsibilities of the safety program manager.*

- Be the organization's risk manager
- Be a member of senior staff with direct access to the chief or director
- Receive and act on recommendations from individuals and the safety committee
- Cause safety and health policies to be developed
- Be liaison with workers' compensation providers
- Be liaison with the department physician on health-related issues
- Investigate injuries or line-of-duty deaths
- Investigate all accidents
- Maintain all records relating to health and safety
- Participate in and develop safety and health training programs
- Perform facility inspection for unsafe conditions

- Evaluate procedures from a safety and health perspective
- Perform cost/benefit analysis
- Attend and participate in postincident analysis
- Have input in to departmental standard operating procedures
- Have input into the specifications of apparatus and equipment to ensure compliance with applicable safety and health standards
- Perform evaluation and analysis of data relating to safety and health
- Maintain awareness of trends in emergency service safety and health including new standards or regulations and court decisions

INCIDENT SAFETY OFFICER

■ Note

Depending on the size and type of the organization, the incident safety officer may be a dedicated position from within the safety division, or may be a first-line supervisor assigned to the safety role on an incident-by-incident basis.

Because most of the injuries that occur to emergency service workers occur at the incident scene, the assignment and function of the incident safety officer is key to the safety and health program. Depending on the size and type of the organization, the incident safety officer may be a dedicated position from within the safety division, or may be a first-line supervisor assigned to the safety role on an incident-by-incident basis. Some departments assign extra units to large incidents and have these crews function in certain safety-related roles, including incident safety officer, accountability, entry control officer, or rapid intervention companies. In some cases, the incident safety officer is in name only, for example, when the assignment is made to anyone who is free at the incident. This is not good practice, because the incident safety officer must have additional knowledge, be well experienced with the incident at hand, and have no other duty. Different incidents may require a different level of safety officer to be assigned (see Figure 9–3). For example, a competent incident safety officer at a fire incident may not be a very good safety officer at a hazardous materials emergency. The incident and the level of expertise of the individuals should dictate the assignment.

NFPA 1521 Standard for Fire Department Safety Officer has requirements and responsibilities of the incident safety officers and requires that one be appointed when the need occurs. The requirements in *NFPA 1521* for the incident safety officer include that that person shall meet Fire Officer I per *NFPA 1021*, have the experience and education to manage incident scene safety for the type of incident, have a knowledge of safety and health hazards involved in emergency operations, know building construction, know the department's personnel accountability system, and understand incident scene rehabilitation.

After arrival at the incident scene, the incident safety officer should evaluate the incident and what is happening. A prediction of what could or is going to happen should also be part of this evaluation. This evaluation involves a 360

Figure 9–3 *An incident safety officer at a vehicle accident.*

degree walk around the scene and examination of available information, such as preincident plans. The incident safety officer should talk to occupants or owners and question them about hazardous situations relating to the property or vehicle, for example, alternative fuel use in a car, or gunpowder in a home, or the storage of explosives. In 1997, a Florida firefighter was killed by a .22 caliber rifle in a house that was on fire and subjected to the intense heat.

Depending on the type of incident, the incident safety officer should assess the operation from a safety point of view and relay findings to the incident commander. The safety officer must be given the authority to instantly stop unsafe acts that are immediately dangerous to responders. The stopping of an assignment must be relayed to the IC promptly as well. During the incident evaluation, the incident safety officer should assess the operating personnel, including the use of proper protective clothing, accountability, and crew intactness. Text Box 9–5 examines some incident types and the role of the incident safety officer. One good technique to remind the safety officer of what to look for at different type of incidents is to provide a clipboard and worksheet that can be completed. These should be durable enough to be used under various weather conditions. There should be a worksheet for each different type of incident that the department may be called on to respond to. Figure 9–4 is an example of a worksheet for multiunit response to a vehicle accident with entrapment. This worksheet can later be used as part of the postincident analysis.

The incident safety officer serves as a member of the IC's command staff and must be a resource for them. The IC must use the safety officer's knowledge and expertise in helping with strategic and tactical decisions.

Fire/EMS Department
Incident Safety Officer Worksheet
Vehicle Accidents

Scene
- ☐ Traffic Controlled
- ☐ Utilities Secured
- ☐ Hazardous
  Materials

Vehicles
- ☐ Stabilized
- ☐ Structural Stability
- ☐ Cargo
- ☐ Fuel Type

Operations
- ☐ IMS
- ☐ Fire Supression
  Support
- ☐ Extrication
- ☐
- ☐
- ☐
- ☐

Personnel
- ☐ Adequate
- ☐ Proper Training
- ☐ PPE
- ☐ Teams Intact
- ☐ Within IMS
- ☐ Accountability
- ☐ Rehabilitation
- ☐

Site Sketch

Units Assigned
_____
_____
_____
_____
_____
_____

Safety Issues
_____
_____
_____
_____
_____
_____
_____

Figure 9–4 *The incident safety officer's checklist for a vehicle accident.*

Text Box 9–5   *Various incident types and the role of the incident safety officer.*

Fire Incidents

- Matches strategy to the situation. Conveys strategy to all operating members.
- Makes sure crews are intact.
- Makes sure IMS is in place.
- Makes sure accountability system is in place.
- Makes sure proper level of protective equipment is being used.
- Determines status of building or vehicle structure.
- Establishes collapse zones and other hazard zones.
- Directs rehabilitation setup.
- Checks physical condition of personnel.
- Continually monitors radio communication.
- Ensures proper scene lighting.
- Communicates unsafe locations such as holes in floors, wires down, and backyard swimming pools to operating teams.
- Makes sure there are means of egress for crews.
- Determines whether risk assessment has been done.
- Determines that rapid intervention crews are ready.
- Makes sure utilities are secured.
- Starts investigation of any on-scene accident or injury.
- Participates in postincident analysis (PIA).

EMS Incidents

- Makes sure IMS is in place if multiple units are operating.
- Makes sure accountability system is in place if needed.
- Determines that protective equipment is available for the task at hand.
- Ascertains structural condition of crashed vehicles, e.g., broken glass, sharp metal.
- Ascertains traffic hazards and creates a buffer zone.
- Reduces exposure to bloodborne products.
- Ensures safe number of rescues, i.e., not doing too much with too little.
- Stops leaking fluid.
- Calls for fire suppression support.
- Initiates infection control.
- Ensures time for rehabilitation.
- Checks physical condition of crews.
- Starts investigation of any on-scene accident or injury.
- Participates in PIA

Hazardous Materials Situation

- Makes sure IMS is in place.
- Makes sure accountability system is in place.
- Makes sure rapid intervention team is available and with the same protective clothing as entry team.
- Ensures teams are operating within level of training and available resources.
- Makes certain teams have adequate protective clothing.
- Establishes proper zones.
- Monitors pre- and postentry physical conditions.
- Ensures that rehabilitation is available.
- Makes sure decontamination process is adequate.
- Starts investigation of any on-scene accident or injury.
- Participates in PIA.

Technical Rescues/Special Operations

- Makes sure IMS is in place.
- Ensures accountability system is in place.
- Ensures there are adequate resources, both human and equipment.
- Makes sure personnel are acting within level of training.
- Ensures protective equipment is available.
- Ensures proper equipment is available—ropes, air supply, personal flotation devices, and others.
- Starts investigation of any on-scene accident or injury.
- Participates in PIA.

THE SAFETY COMMITTEE

One of the most common methods to involve employees and to have a forum for employees to share concerns about the organization's safety and health program is through the use of a safety committee. The safety committee can be a resource for the safety program manager and other safety professionals in the organization. Some safety committees, as a group, review accidents and injuries and assist the program manger with efforts to reduce them. Having employee involvement on the committee helps to deliver to all employees the message that safety is a priority and gives the employees the feeling that they are involved, through representation, in formulating procedures and safety-related policies (see Figure 9–5).

If the members of an organization are represented by a labor union, often there is a contractual requirement to have a safety committee and to have employee representation on it. Even if there is no requirement, the safety committee

■ Note
The safety committee should have representation from all levels of the organization.

Figure 9–5 *The safety committee must have members from all levels of the organization.*

should have representation from all levels of the organization. Do not forget to have all administrative personnel represented, including people in nonresponse roles, such as dispatchers or office staff.

Once formed, the safety committee should elect a chairperson who may or may not be the safety program manager. The committee should meet as often as required to do business, but no less than four times a year. The minutes of the meetings should be published in a format that allows all members of the department to read and review them. Any recommendations from the committee should be forwarded to top management for action. Depending on the organization, the safety committee chairperson may have direct access to the top manager without going through the various levels of command. Text Box 9–6 shows an example of safety committee policy.

Text Box 9–6   *Example of a safety committee policy.*

Safety Committee

This section formally establishes the safety committee, provides a clear statement of duties and responsibilities, and outlines administrative, procedural, and reporting requirements. The safety committee will meet quarterly.

Organization and Duties

The duties and responsibilities of this committee are to:

1. Actively participate in safety and health training programs and evaluate program effectiveness.

2. Review inspections to detect unsafe conditions, materials, practices, and environmental factors.

3. Review and analyze all workplace accidents for hazards, trends, and proper corrective actions and recommend updates to safety rules and procedures as needed.

4. Recommend methods and activities aimed at hazard reduction and/or elimination.

5. Review, compile, and distribute hazard, safety, and health information to the department.

6. Review suggestion or incentive programs.

7. Review new laws, regulations, and guidelines to assess impact and develop courses of action to effectively meet and comply with requirements.

Membership

Participation on the safety committee will include, as a minimum, the following assigned members from our organization:

The safety committee chairperson is _____.
(May be one of the positions below.)

Position	Name of Assigned Member
Supervisor	_____
	_____
Firefighter	_____
	_____
Safety officer	_____

Order of Business

The order of business for the safety committee meeting will be:

1. Record of attendance

2. Approval of previous minutes

3. Unfinished business/open action items

4. Review of accidents, equipment failure, and recalls

5. Presentation of new business

6. Reports on special assignments

7. Reports of inspections

8. Suggestions, awards, recognitions, and incentives

9. Special feature (film, talk, demonstration)

Record Keeping

Formal documentation shall be kept for each meeting. Confidentiality is critical in the discussion of incidents and injuries and must be addressed both in the discussion and the writing of the minutes. Copies of safety committee minutes are to be posted in all work locations and are to be forwarded to the fire chief as well as safety committee members.

SUMMARY

Each member of the organization has a particular responsibility in the safety and health program. From the individual responder's role in maintaining teamwork and following procedures to top-level management appropriating money for safety equipment, the program will not be effective without everyone doing his or her part and assuming his or her role (see Figure 9–6).

Some specific roles that must be considered in the program are the safety program manager, the incident commander, and the incident safety officer. The safety program manager is the nucleus of all activities relating to the program and must be knowledgeable in multiple subjects dealing with safety and risk management. The incident commander also plays a vital role as the person who selects the strategy and tactics for a given situation. Commitment to safety and an understanding of risk management is an important attribute for the incident commander. The incident safety officer responds and becomes a member of the incident commander's staff. The incident safety officer performs an evaluation of the incident and the operations and recommends operational changes to the incident commander. The incident safety officer must have the knowledge and experience specific to the particular incident at hand. There are various safety-related issues that are particular to incident type.

The safety committee is an extension of the organization's membership and safety manager. The safety committee provides a forum in which members make suggestions on safety and health matters. The committee can also be used to investigate accidents and injuries and make recommendations for improving existing or implementing new procedures designed to reduce occupational injures. The safety committee should have membership and representation from all levels of the organization, and may or may not have direct access to the organization's senior manager.

Concluding Thought: Every person in the organization plays an integral part in the overall safety and health program. Everyone has a specific role with defined responsibility.

Figure 9–6 *Like teamwork at an emergency, each member of the organization has an integral role in the safety and health program.*

REVIEW QUESTIONS

1. According to this text the first-line supervisor would be responsible for:

   A. Ensuring the proper protective clothing is worn at a single-unit medical incident.

   B. Communicating safety concerns on the crew's behalf to the next level of management.

   C. Ensuring that safety procedures are followed while responding to an incident.

   D. All of the above are responsibilities of the first-line supervisor.

2. List three good attributes of an individual responder.

3. List three don'ts or things that an individual responder should avoid.

4. Attacking a room and contents fire from the inside of a structure would be considered an offensive fire attack.

   A. True

   B. False

5. A _____ fire attack would be one in which the responders protect exposed structures and do not enter the fire building because of fire conditions.

6. List three things that an incident safety officer might look for at a fire incident.

7. List three things that an incident safety officer might look for at an EMS incident.

8. List three things that an incident safety officer might look for at a hazmat incident.

9. List five roles of the safety program manager.

10. Describe the relationship between the incident commander and the incident safety officer.

ACTIVITIES

1. Compare the activities of your department's safety committee to the recommendations in the text. What could be improved? If you do not have a safety committee, how could one be started?

2. What are the procedures in your department for assignment of an incident safety officer?

   What improvements can be made? How well does the incident safety officer interface with the incident commander?

3. Examine your own roles at incidents. Do you find yourself displaying any of the don't traits listed in the chapter? What can you do yourself to improve the safety of your team?

Chapter
10

SAFETY PROGRAM DEVELOPMENT AND MANAGEMENT

Learning Objectives

Upon completion of this chapter, you should be able to:

- List the essential elements of a safety and health program.
- Describe the process required for the development of goals and objectives.
- Develop an action plan based on the goals and objectives.
- Perform a cost–benefit analysis.
- Describe the relationship of training to the safety and health program.
- Describe the process for developing standard operating procedures.

CASE REVIEW

A small midwestern fire and emergency medical services department realized, through risk identification and analysis, that the number of back sprains and strains occurring within their combination department were well above the national average. This department was just developing a comprehensive safety and health program and was performing the risk identification and analysis during this startup period. The department had appointed a safety program manager and a safety committee.

Both the safety program manager and the safety committee recognized the issue with the back injuries and sought solutions. First, the problem, or risk, was identified, specifically a higher-than-average number of back injuries. The safety program manager and safety committee then developed a broad-based goal of reducing back injuries by 50% the first year and an additional 25% the second year and each year thereafter. There were three goals to the program: The first dealt with training and procedures, the second with placement of equipment on the apparatus, and the third with physical readiness to perform the lifting tasks.

The first objective focused on training the employees to recognize that a problem existed. This was accomplished through a training session in which the local statistics were compared to national statistics with the information provided to all members. Also at this training session, a new procedure was introduced that required any nonemergency lifting of an item weighing over 50 pounds to be done with two people.

The second objective required the committee to look at each piece of equipment and where it was carried on the apparatus. The equipment locations on the apparatus were changed so that the heaviest items were placed at or below waist level of the average employee.

To achieve the third goal, an exercise physiologist was contacted, and exercises particular to strengthening back muscles were added to the daily fitness routine of the members.

After one year the program completely eliminated all back injuries in the department. And, excluding personnel cost during training, the program did not cost the department a penny.

INTRODUCTION

This text has thus far allowed the reader to form a good understanding of what is required for a comprehensive safety and health program. Having the information presented as tools, the text now must lead the reader down the path to getting started. With all this information, it is difficult to determine just where to start.

This chapter provides just that information—what the essential elements are to a successful program, the setting of goals and objectives, action planning, overcoming barriers, performing a cost–benefit analysis, relating the training program to the goals of the safety program, and developing standard operating procedures. Although presented in this order for the purpose of the chapter, depending on your organization, the order of the process may be different.

ESSENTIAL ELEMENTS

The first step in the process of developing the safety and health program is to determine what essential elements are needed. This may vary by department type, size, and nature of responses. In some cases, one role may be shared by more than one person, or one person may perform more than one role. Clearly, using the information presented in this text and applicable standards such as *NFPA 1500*, the following elements are essential to an effective program:

- Top management that is committed to the program
- A safety and health program manager
- A comprehensive risk management plan
- A safety and health policy
- Some type of record-keeping system for data analysis
- Incident safety officer(s)
- A training program
- Standard operating procedures
- Proper equipment and PPE that meets requirements and standards
- A safety committee
- A department physician
- An accident and injury investigation process
- Infection control program
- Provisions for critical incident stress management
- Access to local, state, and national injury and death statistics
- A process to analyze and implement policies to comply with local, state, and national regulations and standards

Safety is not cheap. Some of the elements, such as a department physician can be expensive, whereas getting the commitment of top management is free. However, the program can be developed over a period of time as funding becomes available. It is better to get something started and then seek out the funding for the more expensive components. At minimum, we need the commitment of top management, a safety program manager, a safety committee, a record-keeping system, and access to data in order to get the program off the ground.

SETTING GOALS AND OBJECTIVES OF THE PROGRAM

Once the essential elements of the safety program are in place and the risk identification process is completed, the safety program manager and safety committee start setting goals and objectives for the program. This goal setting is accomplished through the common approach taught in many management courses and may be the organization's strategic planning for safety.

Goals

goals
broad statements
of what needs to
be accomplished

The first step in this process is to determine **goals**. Goals are defined as broad-based statements with a measurable outcome and time frame. Goals can be developed by the safety program manager alone or with the safety committee. However, using the group process and the safety committee will probably result in better acceptance by the members, and, generally, the group process provides a better result. There may be any number of goals set forth and these become the road map to guide the safety program. It should be cautioned that the number of goals developed for a particular time frame should be limited, and the goal should be realistic and obtainable. If there is too much to do in the specified time frame or if goals are never obtained, the organization's members may quickly lose interest and the program may suffer.

Using the case study presented at the beginning of the chapter, the safety committee had identified a particular problem with back injuries. After the problem was identified, the committee began goal development. After discussion, consensus was reached and the following goal recommended.

■ Note

The number of goals developed for a particular time frame should be limited, and the goal should be realistic and obtainable.

> Safety Program Goal 1: To develop a comprehensive back injury reduction program to reduce the number of job-related back injuries by 50% within the first 12 months of implementation with additional 25% reduction each 12 months thereafter.

In analyzing this goal, it is determined that the goal is broad enough, does not give any specifics, is measurable (50% reduction), and has a specified time frame (12 months). This would be a good goal statement for any back injury reduction program.

Once the goal statement has been developed, more specific statements of action must be developed in order to meet the goal. These specific statements, or objectives, provide the road map to reach a goal. Whereas a goal is broad based, an objective is specific, but also must be measurable and within a given time frame.

■ Note

Whereas a goal is broad based, an objective is specific, but also must be measurable and within a given time frame.

Objectives

There is no magic number of objectives needed to reach a goal. Instead, it depends on the complexity of the goal and the number of interrelated activities needed. An objective must only deal with one specific activity and, as with goals, be obtainable within the given time frame. The following is a poorly written objective based on our fictitious back injury problem:

■ Note

An objective must only deal with one specific activity and, as with goals, be obtainable within the given time frame.

> Objective 1.1: A training program will be developed and presented to all employees, and each employee will be issued a back support belt within 30 days.

An analysis of this objective reveals a couple of flaws. First, the objective deals with more than one interrelated issue, training and support belts, and, second, the time frame set forth may not be reasonable. The four following objectives are written to comply with the suggested guideline and are designed to meet the back injury reduction program goal:

> Objective 1.1: Within 90 days, develop a standard operating procedure requiring assistance when performing nonemergency lifting of equipment over 50 pounds and train employees in the application of the procedure.

> Objective 1.2: Within 120 days, form a team to research options for relocating heavy equipment on the apparatus to lower levels and submit a written report with recommendations to fire department management.

> Objective 1.3: Within 120 days, consult with the department physician to develop a fitness routine designed to strengthen back muscles and incorporate education on proper lifting into daily fitness training.

> Objective 1.4: Within 180 days, research the use of back support belts in emergency service applications, including organizational results, and file a written report to the safety committee including recommendations and a cost–benefit analysis.

These objectives are all clear and to the point. They are measurable and include reasonable time frames. Each could be easily analyzed at the end of the specified time frames for completion. See Text Box 10–1 for a summary of the goal and objectives for the example program.

Text Box 10–1 *Summary of goal and objectives for the back injury example.*

Problem: The average frequency of back injuries in the fire department is higher than that in comparable departments.

Safety Program Goal 1: To develop a comprehensive back injury reduction program to reduce the number of job-related back injuries by 50% within the first 12 months of implementation with additional 25% reduction each 12 months thereafter.

Objective 1.1: Within 90 days, develop a standard operating procedure requiring assistance when performing nonemergency lifting of equipment over 50 pounds and train employees in the application of the procedure.

Objective 1.2: Within 120 days, form a team to research options for relocating heavy equipment on the apparatus to lower levels and submit a written report with recommendations to fire department management.

Objective 1.3: Within 120 days, consult with the department physician to develop a fitness routine designed to strengthen back muscles and incorporate education on proper lifting into daily fitness training.

Objective 1.4: Within 180 days, research the use of back support belts in emergency service applications, including organizational results, and file a written report to the safety committee including recommendations and a cost–benefit analysis.

The goal and objective process is dynamic and changing from two perspectives. First, the goals and objectives have to be developed for each problem area and may change over time as new problems are identified. Second, as part of the process, each goal and objective should be reevaluated during the implementation process to see that the objective is being met, and if not, what has to be changed.

Safety program goals and objectives should be published and recognized by all members of the organization. This can be accomplished through the safety committee meeting minutes or any other departmentwide distribution process.

ACTION PLANNING

As described in the previous section, the goal statements are the road map for the safety program and the objectives are the road map to the goal. Then what is the road map to the objectives? The answer is the action plan (see Figure 10–1). The action plan is a step-by-step written guide to meeting an objective. Each objective should have an action plan developed.

The action plan has several components to be considered. The action plan should list the goal and objective that it is designed for and team member's names. The action plan should be developed by the team that has been assigned the objective and should contain very specific step-by-step actions, often set up in tabular form. For each step, the action, a completion time benchmark, the person responsible, the resources needed, any support or roadblocks anticipated, and a completion date should be examined and included. The only column that is sometimes left blank is the support/roadblock column, although generally support and roadblocks should be identified during the action planning process. Table 10–1 is an example action plan for objective 1.2 of our back injury reduction plan.

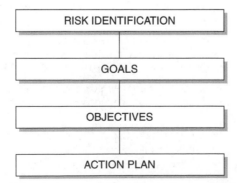

Figure 10–1 *The relationship between goals, objectives, and the action plan.*

Table 10–1   *Sample action plan.*

Safety Committee
ACTION PLAN

Date: January 1, 2004   **Team Members:** Jones, Smith, Big, Little

Goal 1: To develop a comprehensive back injury reduction program to reduce the number of job-related back injuries by 50% within the first 12 months of implementation with additional 25% reduction each 12 months thereafter.

Objective 1.2: Within 120 days, form a team to research options for relocating heavy equipment on the apparatus to lower levels and submit a written report with recommendations to fire department management.

Step No.	Action	Date To Be Completed by	Person Responsible	Resources Needed	Support(S)/ Roadblocks(R)	Date Completed
1	Review SOP adoption guidelines	1/10/04	Jones	None	None	
2	Contact other similar departments for SOPs	1/15/04	Smith	Telephone/E-mail Fax	(S) Chief association/ (S) Statewide labor organization	
3	Develop SOP	1/30/04	Smith	Computer	(S) Safety committee	
4	Determine heavy equipment	1/10/04	Big/Little	None	None	
5	Determine compartmentation options	1/30/04	Big/Little	Vehicle compartment inventory	(S) Company officers	
6	Have SOP draft approved	2/15/04	Smith	None	(S) FD Staff (R) FD Staff	
7	Begin training on SOP	2/20/04	Smith/Jones	Overhead training time	(R) Personnel acceptance	
8	Move equipment as recommended	3/1/04	Big/Little	None	(R) Personnel acceptance	
9	Full implementation	3/30/04	Team	None	None	
10	Evaluate outcome Recommend improvements	3/30/04	Team	None	(S) FD Staff	

As seen in Table 10–1, the action plan covers all of the necessary steps to fully implement the objective. Further progress can easily be measured throughout by using the completed benchmarks. Certain support and roadblocks have been identified so the team can anticipate these issues during the implementation process. This action plan example, although maybe not all inclusive, is a good example of a road map to the objective.

COST–BENEFIT ANALYSIS

Although the action plan presented in Table 10–1 did not include a cost–benefit analysis, very often one is necessary to support a position, particularly if the objective involves any financial outlay. A cost–benefit analysis is often used to show that the initial outlay for a program will save in future reduction of risk. Examples might include the purchase of bunker pants and the reduction in burn injury costs or the cost of an extensive driver training program and the associated reduction in vehicle accidents.

Often when using a cost–benefit analysis, some assumptions must be made and sometimes the data are based on estimations of improvement rather than actual results. Unfortunately it is almost impossible to determine the exact outcome of a program prior to its implementation. However, using information from similar agencies that have done similar projects, an estimation of expected outcomes can be formed. Another way to view outcome expectation is to analyze historical injury data and determine which would likely have been prevented if the program in question had been in effect. Another issue with cost–benefit analysis is that you cannot put a price on some losses that the emergency services incur. For example, the loss of one life is too much, and therefore a program would have to be undertaken regardless of the resultant cost–benefit analysis result.

A cost–benefit analysis typically allows the safety program manager to evaluate the cost-effectiveness of a program. The analysis examines the current cost of the risk and compares those costs to the cost of program implementation, considering both direct and indirect losses and costs. Recall from Chapter 3 that the cost of a risk is measurable and can be both direct and indirect. Direct cost might include the costs of medical treatment, the overtime paid to cover a vacancy on a crew, or the cost of replacing equipment. Indirect costs include loss of productivity, the loss of using the equipment, stress-related concerns of coworkers, and possibly the cost of replacing the employee. Direct and indirect costs may also be applied as program costs.

■ Note

The first step in the cost–benefit analysis is to describe, numerically, the cost of the risk currently.

The first step in the cost–benefit analysis is to describe, numerically, the cost of the risk currently. Again, using our back injury example, a study can be performed and can describe what is occurring at the present time, without intervention. The next step is to determine the cost of the risk after the intervention has been implemented. After this step, the manager can determine whether this measure would be effective from a purely cost standpoint. Finally, the cost of the program implementation has to be calculated. Once these three areas have been calculated, an informed decision can be made. Remember that a program that may only save $1,000 per year and cost $5,000 to implement may still be a good program because of the organizational benefit after the fifth year. Figure 10–2 is an example spreadsheet using the back injury reduction example. The analysis is for illustration only and is not to be construed as showing exact costs.

An examination of Figure 10–2 reveals that the organization would save $12,180 the first year. Aside from some ongoing training and replacement back

Cost/Benefit Analysis	
Goal: Development of Back Injury Prevention Program—First Year	
Current Situation and Costs	
25 Back injuries per year with resultant hospital/doctor visit average $1,000 per visit	
and an average 1.5 (36 hours) days of work lost	
Direct Costs	
Medical expenses	$25,000
Indirect Costs	
Overtime to cover vacancies on shift, average hourly rate $10.00	
36 × 25 × 1.5 × $10.00	$13,500
Total Costs	$38,500
Future Estimation of Situation	
13 Back injuries per year with resultant hospital/doctor visit average $1,000 per visit	
and an average 1.5 (36 hours) days of work lost	
Direct Costs	
Medical expenses	$13,000
Indirect Costs	
Overtime to cover vacancies on shift, average hourly rate $10.00	
36 × 25 × 1.5 × $10.00	$ 7,020
Total Costs	$20,020
First Year Savings (Estimation based on research)	$18,480
Program Costs	
Direct	
Physicians time	$ 2,000
Back supports	$ 2,500
Indirect	
Training time	$ 1,000
Apparatus compartmentation change	$ 800
Total Program Costs	$ 6,300
Total Cost/Benefit—First Year	
First Year Savings minus Program Implementation Costs	$12,180

Figure 10–2   *An example cost–benefit analysis of a back injury prevention program.*

supports, the cost of the program in future years would be greatly decreased. Therefore, the savings over time would be higher and could be calculated using the same format.

TRAINING

There is a direct and close relationship between training and safety (see Figure 10–3). In fact, in many organizations the training officer or division has assumed the safety program functions. Many of the training programs offered have a basis in operating safely. Many of the safety initiatives that a department undertakes are disseminated through the department's training programs. From the back injury program example, a key component involved the training of members.

As presented in Chapter 2, a number of standards and regulations dictate how an emergency service organization operates. Within these standards and regulations are requirements for certain levels of training. A very good example is the hazardous materials waste operations regulation, which bases training requirements on five levels of response. Training mandates are designed in hopes of ensuring that a minimum level of training is provided to responders in order that the responders can handle the incident safely and recognize dangers. Many states require a minimum level of training for firefighters or paramedics. Within these training requirements, one can always find some requirement related to safety, whether it be use of SCBA or infection control.

Training is also important as the organization's safety program develops. The training program can be a vehicle for introducing and testing new procedures.

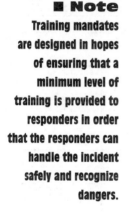

■ Note

Training mandates are designed in hopes of ensuring that a minimum level of training is provided to responders in order that the responders can handle the incident safely and recognize dangers.

Figure 10–3 *A training class on fire extinguishers.*

Remember, the best accountability system on paper that does not work during training evolutions will probably not work at a real incident. Training may also be developed to help deal with an existing injury problem, such as lifting injuries or thermal burns.

More than any other function within an emergency service organization, the training staff should be the most integrated into the safety program, both in development and in implementation.

Vast resources are available for the organization's training program including fire and EMS professional organizations, the United States Fire Administration, state fire/EMS training center, local colleges, universities, the fire academy, textbooks, and the NFPA (see Text Box 10–2). Using the Internet (described in more detail in Chapter 12) can provide training information, ideas, and experiences from other emergency service departments.

Text Box 10–2   *NFPA standards relating to fire department training.*

NFPA 1401 Recommended Practice for Fire Service Training Reports and Records

NFPA 1402 Guide to Building Fire Service Training Centers

NFPA 1403 Standard on Live Fire Training Evolutions

NFPA 1404 Standard for Fire Service Respiratory Protection Training

NFPA 1410 Standard on Training for Initial Emergency Scene Operations

NFPA 1451 Standard for a Fire Service Vehicle Operations Training Program

DEVELOPING STANDARD OPERATING PROCEDURES AND SAFETY POLICIES

Developing safety procedures and policies is necessary in order to meet some of the goals and objectives defined in the program. After development, approval, and implementation, they must be reviewed for effectiveness and updated as necessary. An excellent resource for the development of SOPs is a USFA publication titled "A Guide to Developing Effective Standard Operating Procedures for Fire and EMS Departments." This guide basically describes a development process based on the following four steps:

1. Conduct a needs assessment
2. Develop the SOP
3. Implement the SOP
4. Evaluate the SOP

In starting the process, the committee developing the SOP or policy should review the applicable goal or objective that the policy has to satisfy, or in other words, perform a needs assessment. Other organizations may be contacted for copies of their SOP for the same objective. A gap analysis can also be performed, which simply answers the question of where we are versus where we should be.

Once the needs assessment has been completed, the committee can develop the SOP based on organizational format. It is important that once a draft is developed, it is sent out for comment. It is a good idea to get comments from other members of the department and from various ranks within the organization. If the SOP or policy is specific to one work function or one workplace, it is very important to target the affected groups. This inclusion allows for greater input and help with member buy-in, once adopted. Getting feedback may also tell the committee that the procedure may not work in real application.

Once this entire process has produced an approved new or changed SOP or policy, all members of the organization should be trained in its application. This step is the implementation. This step, which requires realistic training, not just a classroom reading, will tell if the procedure is practical for operations. Further feedback should be provided to ensure the procedure will work.

After the implementation step, there must be evaluation of the SOP, which should include feedback that can be used to update or adjust the SOP as needed. Figure 10–4 graphically shows this process as described in the USFA publication.

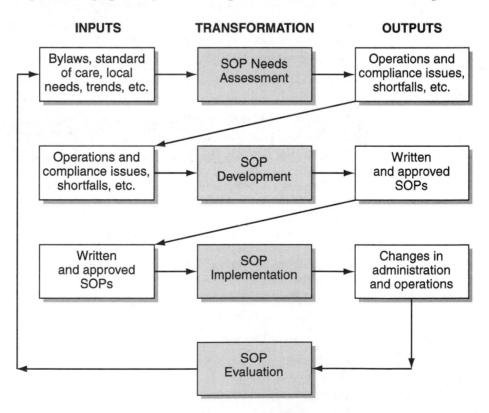

Figure 10–4 *USFA SOP process.* (Courtesy USFA.)

SUMMARY

In order to be effective, the safety and health program must have several essential elements including a top management that is committed to the program, a safety and health program manager, some type of record-keeping system for data analysis, an incident safety officer or officers, a training program, standard operating procedures, proper equipment and PPE that meets requirements and standards, a safety committee, a department physician, and access to local, state, and national injury and death statistics. Some of these items may come after program development. However, at a minimum the program should have the commitment of top management, a safety program manager, a safety committee, a record-keeping system, and access to data.

Once risks are identified and analyzed, the safety program manager, in conjunction with the safety committee, must develop goals and objectives designed to minimize the risks. Goals are broad-based statements that are measurable and realistic and provide a guide to the safety program. Objectives are more specific and are the guide to meeting the goal. There may be one or many objectives for each goal. An action plan is developed based on each objective. The action plan is a step-by-step guide to meeting the objective. For each step in the action plan—the action, a completion time benchmark, the person responsible, the resources needed, any support or roadblocks antici-

pated, and a completed-by date—should be examined and included.

For most goals and objectives, a cost–benefit analysis should be performed. The cost–benefit analysis examines the current situation in terms of direct and indirect cost. The costs of the risk after program implementation and the expected cost of the program are compared to the current cost, so that an informed decision can be made regarding implementation.

Training and safety are very closely related and somewhat dependent on each other. In some organizations the training division actually does the safety function as well. Many training programs are designed for educating the members on how to perform tasks safely. Development of safety- and health-related SOPs and policies are also integral to the program. These can be developed by the safety committee after gap analysis, review of the safety goal and objectives, and after contact with other organizations with similar policies. Once a draft is developed, it is important to get feedback from all levels of the department and to train all members on the implementation.

Concluding Thought: The safety and health program has several components. Using these components, the safety program manager goes through various processes to implement the program.

REVIEW QUESTIONS

1. A department physician is an essential component to have in place before a safety program can be established?

   A. True

   B. False

2. A goal is a broad statement with measurable results?

   A. True

   B. False

3. An action plan can be considered the road map to a goal.

   A. True

   B. False

4. List the four-step process for developing a SOP as outlined in the USFA publication "A Guide to Developing Effective Standard Operating Procedures for Fire and EMS Departments."

5. Which of the following is applicable to a cost–benefit analysis?

    A. The cost of the safety gear to be purchased.

    B. The current direct cost of the injuries if the program were not introduced.

    C. The indirect costs associated with the program after implementation.

    D. All of the above information would be necessary.

ACTIVITIES

1. Select an injury problem within your department. Write one goal and at least two objectives to develop a program to reduce or eliminate the problem.

2. Write an action plan for each objective.

3. Perform a cost–benefit analysis on the program.

4. Review your department's safety program and the relationship to the training division. Are they working together? Do they have similar goals? Is it effective?

5. Review your department's SOP development process. How does it compare to that presented in the text?

Chapter

11

SAFETY PROGRAM EVALUATION

Learning Objectives

Upon completion of this chapter, you should be able to:

- Describe the purpose for evaluation of the health and safety program.
- Compare the two types of evaluations, process and outcome.
- Explain who has the responsibility for evaluation.
- Describe a recommended frequency for evaluation and the factors that affect the frequency.

CASE REVIEW

The large urban EMS agency had placed a comprehensive injury reduction program into effect more than 3 years ago. The program was evaluated annually using outcome evaluation. This evaluation process examined such areas as injury rate and severity, attitude changes among personnel, a knowledge change in the personnel regarding injury prevention, and an evaluation of how well the policies were being complied with.

The first annual evaluation revealed that the injury reduction program was, in fact, meeting its goal of reducing injuries. In the first 12 months, the injury rate dropped by 50%. Subsequent annual evaluations showed similar changes in the injury rate, however injury severity was shown to be on the increase. The outcome evaluation focusing on the foregoing areas revealed that during the first years of the program, the policy changes were complied with very well throughout the department. However, due to a number of new employees hired over the subsequent 2 years and a number of promotions, many of the policies associated with the program lacked priority in terms of compliance.

Although the number of injuries continued to decrease, the severity increased as employees were failing to follow established procedures. It is imperative in program evaluation that all components be evaluated from several different dimensions.

INTRODUCTION

No health and safety program is complete without some sort of evaluation process. In fact, *NFPA 1500* Chapter 4 requires that the risk management plan be evaluated. Once the program has been designed and implemented, a process must be undertaken to evaluate its effectiveness. This evaluation may take any number of forms, depending on the goal or objectives being evaluated. Furthermore, the time frame for evaluation will also be goal dependent. Simply put, the method of evaluation should compare where we are now to where we were prior to program implementation. In order to be effective, an understanding of the evaluation process and the measuring of results must be undertaken. Once this knowledge is gained, the process can be included in the program (see Figure 11–1).

■ **Note**
The method of evaluation should compare where we are now to where we were prior to program implementation.

THE EVALUATION PROCESS

The process of evaluation may take different forms; however, it usually involves the comparison of statistics from one time period with another time period. The results can then be compared to what was expected. Differences between expected outcomes and actual outcomes should be evaluated for program changes. There are three reasons to evaluate the program:

1. To see if the program is effective
2. To determine the response to the program from the members' perspective
3. To facilitate program changes

Figure 11–1 *A safety program manager analyzes data for the annual evaluation.*

process evaluation

the evaluation of the various processes associated with a program or task that is ongoing

outcome evaluation

an evaluation that answers the question, "Did the program meet the expected goals?"

The are two methods of evaluation that should be used: **process evaluation** and **outcome evaluation**.

Process Evaluation

Process evaluation is an analysis of the procedures of the program and is undertaken throughout the program. Basically, the process evaluation answers the question, "How well did the processes in the program do what they were intended to do?" For example, the back injury prevention program designed in Chapter 10 had a component involving the training on proper lifting techniques. A process evaluation for this component would examine whether the training had the desired effect. Simply put, are the responders using the proper lifting techniques they were trained to do? The following questions are necessary in the process evaluation:

- Who was affected by the program?
- To what extent were they affected?
- Are improvements occurring as planned?
- What part of the program appears to be most effective?
- Which parts of the program appear to be least effective?

There are a number of ways to perform the process evaluation including written or practical testing on safety policies and procedures, visual observation at incident scenes and around the station, and feedback from supervisors. Process evaluation can be used to measure results against program objectives. Process evaluation will tell the safety program manager what program changes are needed to increase the effectiveness of the program.

Outcome Evaluation

Outcome evaluation is measured after the program has been in effect for a while. This evaluation examines what the program did. The outcome evaluation answers the question of whether the program is being effective. Again using the back injury prevention program as an example, an outcome evaluation would be used at the end of a time period, probably a year, to determine what effect the program had. Did the program actually reduce back injuries? Several areas should be analyzed, including:

- Examining the current injury rates and severities, comparing them to the rates prior to program implementation (using the goals and objectives as benchmarks)
- Measuring the change in knowledge, behavior, and performance
- Measuring behavior changes
- Analyzing the changes in the physical environment
- Measuring the response to policy changes.

Much of the information needed can come directly from the injury statistics, however, there will be other required measurements. For example, when analyzing the change in attitude and knowledge a more **cognitive** measurement may have to be used. Outcome evaluation is often used to measure the results obtained against the overall program goal.

cognitive
skills learned through a mental learning process as opposed to practical learning

The safety and health program evaluation should include both of these evaluation methodologies. While for the purpose of this textbook, these examples are simplistic, the reality is that the complete safety and health program must be evaluated. This evaluation can be as a whole or as individual parts, as long as the evaluation results in a comprehensive look at the program.

RESPONSIBILITY FOR EVALUATION

Who has the responsibility for program evaluation? The simple answer is that everyone has some responsibility for it. Everyone can and should play some role in the evaluation process. The fire chief or EMS director has the ultimate responsibility to see that program evaluation is done, but the safety program manager may oversee the process. The responders perform a process evaluation each time they respond to an alarm. The evaluation may be accomplished through the safety committee, incident safety officers, senior staff members, and first line supervisors as well.

The safety committee can assist with compiling and analyzing the injury data and also help with the determination, through feedback, of knowledge, performance, and behavior changes. The incident safety officers are valuable in providing feedback during process evaluation and for measuring compliance with new policies or procedures. Senior management is part of the process by making resources and data available to the safety and health program manager and the safety committee. First-line superiors also are feedback providers and can very often measure knowledge and attitude changes. First-line supervisors also are integral for feedback on policies and procedure effectiveness. Finally, the individual member should be empowered to make suggestions to improve program components and can measure peer attitude and knowledge changes.

Each of the program evaluators discussed so far has been internal to the organization; however, the *NFPA 1500* Standard recommends external evaluations every 3 years. Having an external evaluation performed is a positive effort for both the safety program manager and the program in general and is highly recommended. Very often a person from the outside who is not close to the issues will find areas in which improvements can be made that may not be as oblivious to the internal evaluators. Performing an external evaluation does not have to be a costly endeavor. Workers' compensation and other insurers may have risk managers on staff and often make these experts available to a client for an external look at the safety program as they also have a stake in the safety program. Personnel such as safety program managers in other local departments may be willing to evaluate the program, as might other fire/EMS chiefs. Local colleges or universities may have a safety and health curriculum and may allow student interns to review your program as a class assignment. Finally, if you want to spend some money, the organization could hire a consultant in the safety and health profession to perform the evaluation.

EVALUATION FREQUENCY

When the program should be evaluated is another question. As mentioned previously, the two types of program evaluation, process and outcome, can help to answer this question. Process evaluation can almost be a continuous process. Outcome evaluation should occur at the end of a specific period of time, usually one year or whenever the program goal or objective defines as a measured time frame.

The frequency of evaluation should be considered dynamic. For example, if a serious injury, or near miss, or a death occurs, an immediate evaluation or analysis into the safety program components dealing with the cause should be undertaken. Findings during a PIA may result in the need to evaluate the program. If new technology is introduced into the profession, an analysis of its applicability to the program should be examined. If there are changes to applicable regulations or standards, the program must be evaluated for compliance.

Program evaluation is a required component of the safety and health program. In fact, program evaluation actually occurs before the program starts; risk identification and analysis are types of evaluations. They just happen to be evaluations of a program before a program begins.

SUMMARY

There are two basic formats for the evaluation of the program, the process evaluation and the outcome evaluation. The process evaluation should be viewed as an ongoing analysis into the program to determine whether the program is reaching the intended recipients and to what degree. It also should reveal what changes have occurred in knowledge, performance, and behavior. The outcome evaluation should occur after the program has been in effect for some period of time and actually measures results. These results can be compared to the program goals and objectives and the program effectiveness can be determined.

Everyone in the organization has some role in the evaluation process. Although the responsibility rests with the safety and health program manager, the safety committee, incident safety officers, senior staff, first-line supervisors, and the members all have a role in the evaluation process. External evaluations are recommended in order to have the program evaluated from a different perspective.

Although a process evaluation is ongoing and the outcome evaluation is after a defined period of time, the frequency of evaluation is dynamic. If a major event occurs in the organization or if new technologies are introduced, a program evaluation is warranted at that time. After the evaluation, regardless of the type or the frequency, the information gained should be used for program enhancement and improvement.

Concluding Thought: A safety and health program cannot be successful without evaluation methods as part of the process.

REVIEW QUESTIONS

1. List the two types of program evaluations.
2. List the components required for each of the two program evaluations listed in Question 1.
3. List three purposes for program evaluation.
4. What is the role of the first-line supervisor in program evaluation?
5. Describe considerations for determining the frequency for evaluation.

ACTIVITIES

1. Compare the evaluation of the safety program in your organization to that presented in the text. How does your process compare?
2. What changes do you recommend?
3. Is the role of the various members of your organization similar to the recommendation of the text in terms of program evaluation? If so, is it effective? If not, is it effective?

Chapter
12

INFORMATION MANAGEMENT

Learning Objectives

Upon completion of this chapter, you should be able to:

■ Describe the purpose of data collection and reporting.

■ Identify the data that should be collected within the organization.

■ Identify the data that should be collected for outside organizations.

■ Describe the purpose and process for publishing an internal health and safety report.

■ Describe the use of the Internet as a health and safety information source.

CASE REVIEW

The safety and health program manager was new in the job, having been assigned only 2 months before. The emergency services director sent a memo outlining concerns about bloodborne pathogen exposures. The director felt that the current procedures and policies within the department were not adequate for the protection of the employees.

The safety program manager sought to examine the current system, review the national experience, and find up-to-date information from the electronic media. First, the exposure records for the organization were analyzed. This was accomplished by doing a computer search of the injury database, searching for records that listed exposure for type of injury. The manager then contacted the IAFF and requested a copy of its most recent injury and exposure report. The information from these two reports were compared and the determination made that the organization's level of exposure was greater than the national average.

Having information in hand, the safety program manager filed a report with the director that explained where the organization was and where it should be. The director gave the OK to develop and implement changes to the existing program.

Wanting to know more about the problem and to look at how other agencies were handling the situation, the safety program manager accessed the Internet. Performing a search on the words *bloodborne*, *pathogen*, and *exposure*, the Internet search engine provided the manager with more than 100 sites to visit that would have information on exposures. Some of these were the Department of Labor, which provided a copy of the OSHA standard on bloodborne pathogens; the Centers for Disease Control and Prevention, which had a copy of the Ryan White law and statistics dealing with the disease; and the USFA, which had documents and sample policies from numerous fire departments across the country.

With the internal and external data available and the information from the Internet, the safety program manager had the resources necessary to develop improvements and implement a new plan.

INTRODUCTION

■ **Note**

In order to be useful, data must be retrievable in a format that is compatible with its intended use.

Like program evaluation, data collection and information analysis is a critical component of the safety and health program. Data collection and analysis is used for many of the other program components discussed, including the development of goals and objectives and the evaluation of the program or components. In order to be useful, data must be retrievable in a format that is compatible with its intended use. Although not always possible, this process should be facilitated through the use of a computer.

Data is collected for use within the organization. Data is also collected for and by outside agencies and used for various other purposes. The collection and analyzation by outside agencies is helpful for local comparison to national trends. Once the internal data is collected and analyzed, an annual health and safety report should be published for all levels of the organization.

The environment is rapidly changing and so is the speed in which we can communicate. Our access to information has never been greater. The safety and health program manager can use the Internet to obtain volumes of information regarding safety and health, regulations and standards, and copies of policies from other similar departments regarding the safety and health program.

PURPOSE OF DATA COLLECTION AND REPORTING

There are many reasons for performing data collection. For starters, data collection and reporting is required for many jurisdictions by OSHA or a state department of labor. It is equally important for the organization to have the information available for analysis and evaluation. Furthermore, good record keeping helps to protect the organization and the members from a legal standpoint (see Figure 12–1).

Records regarding workers' compensation and other insurance claims can help predict increases or decreases in premiums that will have future budget impacts. Further, the mandatory reporting requirements of federal and state agencies usually carry fines associated with noncompliance.

Bear in mind that most, if not all, medical-related data and information is considered confidential and the access to the information should be kept to only those with a need to know. The first-ever federal privacy standards to protect patients' medical records and other health information provided to health plans, doctors, hospitals, and other health care providers took effect on April 14, 2003. This standard provided patient privacy protections as part of the Health Insurance Portability and Accountability Act of 1996 (HIPAA). This act probably will have some impact on the medical data collected as part of the safety and health program, depending on the type of organization. Therefore, it is important during the development of evaluation of the safety and health program that legal advice be obtained regarding injury reporting and privacy issues. After seeking legal direction, department SOPs must be developed and implemented that include who has access to medical-related data and why.

INTERNAL DATA COLLECTION

Internal data collection is that associated with health and safety data generated from within the organization and is used within the organization. Several common data types fall into this category, including:

- Injury reports
- Accident reports
- Individual medical histories
- Drug-free workplace test results
- Reports dealing with an employee killed in the line of duty
- Exposure records

SOUTHWEST FLORIDA PROFESSIONAL FIREFIGHTERS
LOCAL 1826 / I.A.F.F.
EXPOSURE REPORT

1. Firefighters Name ______ 2. Social Security # ______ 3. 1826 Dist. # ______
4. Incident Date ______ 5. Alarm Time ______ 6. Incident # ______

A MEDICAL / RESCUE EXPOSURE

7. Patients Name ______
8. Facility Patient Transported To ______
9. Transporting Agency ______
10. Name of Suspected or Known Disease: ______
11. Precautions Taken By You at Scene and After Exposure
☐ gloves ☐ face mask ☐ goggles ☐ protective gown
☐ washed exposed skin ☐ physician treatment (if this is checked complete *section F*)

How you were exposed	A. Topical Skin Contact	B. Needle Puncture	C. Direct contact to open wound	D. In Eyes	E. In nose or mouth
12. BLOOD					
13. URINE					
14. SPUTUM					
15. VOMITUS					
16. FECES					
17. OTHER					

B *FIRE INCIDENT TYPE*

18. ☐ Residental Fire 19. ☐ Industrial Fire 20. ☐ Vehicle Fire 21. ☐ Commercial Fire 22. ☐ Wildland Fire 23. ☐ Trash/Dumpster fire 24. ☐ Marine Fire 25. ☐ Explosion 26. ☐ Rescue 27. ☐ Spill
28. Other (describe in 1 or 2 words) ______ 29. Type of occupancy (1 or 2 words) ______

C LENGTH OF EXPOSURE BY FIRE STAGE / ACTIVITY

FIRE STAGE :	A <1 hr.	B 1-2 hr	C 2-3 hr	D 3 + hrs
30. Incipient				
31. Free Burning				
32. Smoldering				
33. Non-Fire Incident				

	A < 1 hr	B 1-2 hr	C 2-3 hr	D 3 + hrs
34. Extinguishment				
35. Empty / Ventilation				
36. Rescue				
36. Light Overhaul				
37. Heavy overhaul				
38. E.M.S.				
39. Investigation				

D SMOKE / CHEMICAL/ OTHER EXPOSURE

41. Smoke Conditions ☐ light ☐ Heavy ☐ None
42. Smoke Color ______

Chemicals Present	A Vapor Gas	B Dust	C Liquid Heavy Mist	D Light Mist	E Smoke	F Solid Powder
43.						
44.						
45.						
46.						

47. Type of Chemical Exposure to Above: ☐ Inhaled ☐ Ingested ☐ Skin Contact ☐ Eye Contact ☐ Other ______

E SYMPTONS

	AT INCIDENT	SYMPTON	AFTER INCIDENT	FOR HOW LONG
48.		Eyes Burn		
49.		Cough		
50.		Nose Bleed / Cough Blood		
51.		Nose/Lung Irritation		
52.		Nausea/Queasiness		
53.		Dizzy		
54.		Ears Ringing		
55.		Headache		
56.		Skin Rash/Burn		
57.		Unconscious		
58.		Other:		

F MEDICAL DIAGNOSIS

59. DID YOU RECEIVE MEDICAL EVALUATION OR TREATMENT AFTER EXPOSURE? ☐ YES ☐ NO
60. DIAGNOSIS ______
61. DOCTORS NAME ______ 62. FACILITY ______

G SPECIAL EQUIPMENT / DECONTAMINATION ☐ PASS ALARMS ☐ BUNKER PANTS ☐ HELMETS ☐ BUNKER COATS ☐ GLOVES ☐ HOODS ☐ SCBA'S ☐ BOOTS

63. WERE YOU SUPPLIED SPECIAL EQUIPMENT FOR THIS INCIDENT? ☐ YES ☐ NO DESCRIBE ______
64. WERE SPECIAL DECONTAMINATION PROCEDURES FOLLOWED AFTER THE EXPOSURE ☐ YES ☐ NO DESCRIBE ______

H CO-WORKERS AT TIME OF EXPOSURE

65. PLEASE LIST NAMES OF OTHER FIREFIGHTERS WORKING CLOSE TO YOU AT TIME OF EXPOSURE ______

I ADDITIONAL INFORMATION (YOUR FUNCTION AT INCIDENT, UNUSUAL CIRCUMSTANCES, ETC.)

66. ______

J 67. DATE REPORT FILLED OUT ______ 68. NAME OF PERSON FILLING OUT REPORT (PRINT) ______

69. SIGNATURE ______

Figure 12–1 *An example of an exposure-reporting form.* (Used by permission of Local 1826 IAFF.)

In order to collect the data required for these reports, standardized forms are often used. The data should then be placed into a computer database for storage and later retrieval. Safety and health program managers should check their local, state, and federal laws for requirements in keeping hard copies of these records

Figure 12–2 *Sample accident/injury-reporting form.* (Courtesy Palm Harbor Fire Rescue.)

Accident / Injury / Investigation Report
PRINT ONLY

Report #_____

TO BE FILLED OUT BY INJURED PARTY   Today's Date:_____

Name of Injured: Sex: Male ❑   Female ❑   Age:

Date & Time of Accident: AM   PM   Location of Accident:

Describe the accident and how it occurred:

Was accident reported to immediate supervisor? Yes ❑ No ❑ | Date & Time reported to Supervisor:

Were there any injuries?   Yes ❑ No ❑   If yes, describe injuries:

Was medical treatment given? Yes ❑ No ❑   If yes, give physician name & address:

Was time lost from work? Yes ❑ No ❑   If yes, explain.

Were department SOP's adhered to? Yes ❑ No ❑   If no, explain.

Were safety standards adhered to? Yes ❑ No ❑   If no, explain:

Was law enforcement notified? Yes ❑ No ❑   If yes, which law enforcement agency?

What could have been done to avoid this accident?

SIGNATURE OF INJURED PERSON:

TO BE FILLED OUT BY IMMEDIATE SUPERVISOR

Cause of accident:

Witnesses to accident? Yes ❑ No ❑   If yes, list witnesses

Was personal protective equipment required? Yes ❑ No ❑   Was it being used? Yes ❑ No ❑   If NO, explain:

Interim corrective actions taken to prevent recurrence:

Permanent corrective action recommended to prevent recurrence:

SUPERVISOR SIGNATURE   DATE:

Status and follow-up action taken by Safety Officer

SAFETY OFFICER SIGNATURE   DATE: REVIEWED BY CHIEF / DATE

and for the length of time they must be maintained. Regardless of the requirements, the information should still be filed electronically, in addition to required hard copies for easy retrieval and analysis. Figure 12–2 is an example of an accident/ injury reporting form with the generally required information.

EXTERNAL DATA COLLECTION

External data is the information, collected internally, that is to be used by another agency outside of the organization. This agency could be state and national databases, workers' compensation carriers, or insurance companies. As might be recalled from Chapter 1, the IAFF/IAFC joint labor management wellness/fitness initiative requires reporting to a national database accessible by local organizations.

The information required for the external agencies may or may not be the same as that needed within the organization. However, usually the internal use forms provide adequate background information for the external reporting requirements.

Workers' Compensation

Clearly, the organization's workers' compensation carrier is one external organization that wants injury data. Generally, there are rules and laws that provide for specific time frames in which a workers' compensation carrier must be notified. The workers' compensation carrier will also, upon request, provide the organization with a summary of claims over a requested period of time. This can be a useful tool in program evaluation. For example, knowing that internally the number of workers' compensation claims have gone down is good news. However, if the cost of the claims, in terms of lost time and treatment, have increased, the program will have to be reevaluated and changes implemented.

Occupational Safety and Health Administration

OSHA 300
log of work-related injuries and illnesses

OSHA 301
the injury and illness incident report

OSHA 300-A
summary of work-related injuries and illnesses

Although OSHA does not apply to every public fire department in the country, there are reporting requirements in the states in which OSHA does apply. The OSHA 1904 regulation requires that the **OSHA 300** log (see Figure 12–3 on pages 210 and 211) and **OSHA 301** incident report be completed within seven (7) calendar days of receiving information that a recordable injury or illness has occurred. All of the entries on the OSHA 300 log must be summarized on the **OSHA 300-A** log at the end of the year. OSHA regulations do provide for the privacy of the injured party. The safety and health program manager should refer to OSHA 1904 for these provisions.

However, there are guidelines on what types of injuries must be reported. An injury must be reported if it results in any of the following: death, days away from work, restricted work or transfer to another job, medical treatment beyond first aid, or loss of consciousness. You must also consider a case to meet the general recording criteria if it involves a significant injury or illness diagnosed by a physician or other licensed health care professional, even if it does not result in death, days away from work, restricted work or job transfer, medical treatment beyond first aid, or loss of consciousness.

Organizations that do not fall under OSHA for public emergency response agencies should contact its state department of labor to determine a state requirement for similar reporting.

National Fire Protection Association

As described in Chapter 1, the NFPA publishes an annual injury report and an annual firefighter fatality report. In order for these reports to reflect a national experience, organizations are asked to participate in the NFPA data collection program. Remember that the injury portion of the NFPA's reports are a sampling designed to predict the national experience.

United States Fire Administration

The USFA also does health and safety data collection thorough NFIRS. Although voluntary, the NFIRS has components for reporting firefighter casualties, both fatalities and injuries. This information is reduced to an annual report that addresses the nationwide experience for those organizations that participate.

International Association of Fire Fighters

Like the NFPA and OSHA, the IAFF collects data for annual injury, exposure, and fatalities reports. The data collected by the IAFF is only from paid fire departments with IAFF affiliation and includes the United States and Canada. The published report considers EMS-related activities to a greater degree than the previous two reports.

National Institute for Occupational Safety and Health

NIOSH began a project in 1997 in which it investigates firefighter line of duty deaths. The project, called Fire Fighter Fatality Investigation and Prevention Program was funded in 1998. Text Box 12–1 describes this program.

Text Box 12–1 *Description of the NIOSH Fire Fighter Fatality Investigation and Prevention Program* (Courtesy NIOSH).

The United States currently depends on approximately 1.2 million firefighters to protect its citizens and property from losses caused by fire. Of these firefighters, approximately 210,000 are career/paid and approximately 1 million are volunteers. The NFPA and the USFA estimate that on average, 105 firefighters die in the line of duty each year.

In fiscal year 1998, Congress recognized the need for further efforts to address the continuing national problem of occupational firefighter fatalities and funded NIOSH to conduct independent investigations of firefighter line-of-duty deaths.

Firefighter Fatality Investigations

The NIOSH Fire Fighter Fatality Investigation and Prevention Program conducts investigations of firefighter line-of-duty deaths to formulate recommendations for preventing future

(*continued on page 212*)

OSHA's Form 300 (Rev. 01/2004)

Log of Work-Related Injuries and Illnesses

You must record information about every work-related death and about every work-related injury or illness that involves loss of consciousness, restricted work activity or job transfer, days away from work, or medical treatment beyond first aid. You must also record significant work- related injuries and illnesses that are diagnosed by a physician or licensed health care professional. You must also record work- related injuries and illnesses that meet any of the specific recording criteria listed in 29 CFR Part 1904.8 through 1904.12. Feel free to use two lines for a single case if you need to. You must complete an Injury and Illness Incident Report (OSHA Form 301) or equivalent form for each injury or illness recorded on this form. If you're not sure whether a case is recordable, call your local OSHA office for help.

Identify the person			**Describe the case**		
(A) Case no.	(B) Employee's name	(C) Job title (*e.g., Welder*)	(D) Date of injury or onset of illness	(E) Where the event occurred (*e.g., Loading dock north end*)	(F) Describe injury or illness, parts of body affected, and object/substance that directly injured or made person ill (*e.g., Second degree burns on right forearm from acetylene torch*)
			/ month/day		
			/ month/day		
			/ month/day		
			/ month/day		
			/ month/day		
			/ month/day		
			/ month/day		
			/ month/day		
			/ month/day		
			/ month/day		
			/ month/day		
			/ month/day		

Public reporting burden for this collection of information is estimated to average 14 minutes per response, including time to review the instructions, search and gather the data needed, and complete and review the collection of information. Persons are not required to respond to the collection of information unless it displays a currently valid OMB control number. If you have any comments about these estimates or any other aspects of this data collection, contact: US Department of Labor, OSHA Office of Statistical Analysis, Room N-3644, 200 Constitution Avenue, NW, Washington, DC 20210. Do not send the completed forms to this office.

Figure 12–3   *OSHA 300 Log.*

Attention: This form contains information relating to employee health and must be used in a manner that protects the confidentiality of employees to the extent possible while the information is being used for occupational safety and health purposes.

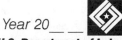

Year 20____ ____

U.S. Department of Labor
Occupational Safety and Health Administration

Form approved OMB no. 1218-0176

Establishment name _____

City _____   State _____

Classify the case

CHECK ONLY ONE box for each case based on the most serious outcome for that case:

Enter the number of days the injured or ill worker was:

Check the "Injury" column or choose one type of illness:

		Remained at Work		Away from work	On job transfer or restriction	(M) Injury	Skin disorder	Respiratory condition	Poisoning	Hearing loss	All other illnesses
Death	Days away from work	Job transfer or restriction	Other recordable cases								
(G)	(H)	(I)	(J)	(K)	(L)	(1)	(2)	(3)	(4)	(5)	(6)
☐	☐	☐	☐	___ days	___ days	☐	☐	☐	☐	☐	☐
☐	☐	☐	☐	___ days	___ days	☐	☐	☐	☐	☐	☐
☐	☐	☐	☐	___ days	___ days	☐	☐	☐	☐	☐	☐
☐	☐	☐	☐	___ days	___ days	☐	☐	☐	☐	☐	☐
☐	☐	☐	☐	___ days	___ days	☐	☐	☐	☐	☐	☐
☐	☐	☐	☐	___ days	___ days	☐	☐	☐	☐	☐	☐
☐	☐	☐	☐	___ days	___ days	☐	☐	☐	☐	☐	☐
☐	☐	☐	☐	___ days	___ days	☐	☐	☐	☐	☐	☐
☐	☐	☐	☐	___ days	___ days	☐	☐	☐	☐	☐	☐
☐	☐	☐	☐	___ days	___ days	☐	☐	☐	☐	☐	☐
☐	☐	☐	☐	___ days	___ days	☐	☐	☐	☐	☐	☐
☐	☐	☐	☐	___ days	___ days	☐	☐	☐	☐	☐	☐
☐	☐	☐	☐	___ days	___ days	☐	☐	☐	☐	☐	☐

Page totals▶ ___ ___ ___ ___   ___ ___

Be sure to transfer these totals to the Summary page (Form 300A) before you post it.

Page ___ of ___

Injury	Skin disorder	Respiratory condition	Poisoning	Hearing loss	All other illnesses
(1)	(2)	(3)	(4)	(5)	(6)

deaths and injuries. The program does not seek to determine fault or place blame on fire departments or individual firefighters, but to learn from these tragic events and prevent future similar events.

The goals of the program are the following:

- To better define the magnitude and characteristics of line-of-duty deaths among firefighters
- To develop recommendations for the prevention of deaths and injuries
- To disseminate prevention strategies to the fire service

Traumatic Injury Deaths

The program uses the Fatality Assessment and Control Evaluation (FACE) model to conduct investigations of fireground and nonfireground fatal injuries resulting from a variety of circumstances such as motor vehicle incidents, burns, being struck by objects, falls, diving incidents, and electrocutions. NIOSH staff also conducts investigations of selected nonfatal injury events. Each investigation results in a report summarizing the incident and includes recommendations for preventing future similar events. Personal and fire department identifiers are not included in the NIOSH investigative reports.

NIOSH staff with respirator expertise also assist with investigations in which the function of respiratory protective equipment may have been a factor in the incident. NIOSH staff evaluate the performance of the SCBA as a system and will conduct evaluations of SCBA maintenance programs on request.

Cardiovascular Disease Deaths

NFPA data show that heart attacks are the most common type of line-of-duty deaths for firefighters. NIOSH investigations of these fatalities include assessing the contribution of personal and workplace factors. Personal factors include identifying individual risk factors for coronary artery disease. The workplace evaluation includes the following assessments:

- Estimating the immediate physical demands placed on the firefighter
- Estimating the firefighters' acute exposure to hazardous chemicals
- Assessing efforts by the fire department to screen for coronary artery disease risk factors
- Assessing efforts by the fire department to develop fitness and wellness programs

Database

Another component of the program is the research database containing information for each injury incident. The database serves as a valuable tool to identify trends and analyze risk factors among line-of-duty injury deaths. Used in conjunction with individual incident reports, the database helps provide valuable information for developing broad-based recommendations for firefighter injury prevention programs. As with the investigation reports, personal and fire department identifiers are not included in the database.

Information Dissemination

Information dissemination is one of the important goals for the NIOSH Fire Fighter Fatality Investigation and Prevention Program. The Fire Fighter Fatality Investigation and Prevention

Program disseminates the investigative reports and other related publications to stakeholders who can take action to help prevent firefighter line-of-duty deaths and injuries. These stakeholders include fire departments, firefighters, program planners, and researchers.

What to Expect During a NIOSH Investigation

NIOSH is notified of a line-of-duty death in a number of ways, including notification by the USFA, a fire department representative, the IAFF, or the state fire marshal's office. NIOSH conducts investigations of both career and volunteer firefighter line-of-duty deaths.

Once notified of a line-of-duty death, a NIOSH representative contacts the fire department to make the necessary arrangements to conduct the investigation. NIOSH investigators review all applicable documents (for example, department SOPs, dispatch records, the victim's training records, coroner/medical examiner's reports, death certificates, blueprints of the structure, police reports, photographs, and video). Additionally, investigators interview fire department personnel and firefighters who were on the scene at the time of the incident. NIOSH also works closely with other investigating agencies. When needed, NIOSH will enlist the assistance of other experts, such as experts in motor vehicle incident reconstruction or fire growth modeling.

Once the investigation is complete, NIOSH summarizes the sequence of events related to the incident and prepares a draft report. Each department and union representative (if applicable) has the opportunity to review this portion of the report in draft form to ensure its technical accuracy. The report is then finalized with the addition of recommendations for preventing future deaths and injuries under similar circumstances. Once the fire department and union (if applicable) have received the final copy of the NIOSH report, it is made available to the public by posting the report on the NIOSH website.

Who Do I Contact for Further Information?

If you have any questions regarding the NIOSH Fire Fighter Fatality Investigation and Prevention Program, contact the NIOSH Division of Safety Research at:

National Institute for Occupational Safety and Health
Division of Safety Research
Surveillance and Field Investigations Branch
1095 Willowdale Road, M/S H-1808
Morgantown, WV 26505-2888
Phone: (304) 285-5916 FAX: (304) 285-5774

PUBLISHING THE HEALTH AND SAFETY REPORT

Once a data collection and retrieval system is put into place, the measurable data must be turned into usable information. Normally, this requires some sort of analysis and the compiling of a report over time, usually annually (see Figure 12–4). Once published, this report should be distributed to all levels of the organization, including elected officials and senior city or county officials. This report can help as an educational tool when support for resources is needed to enhance the

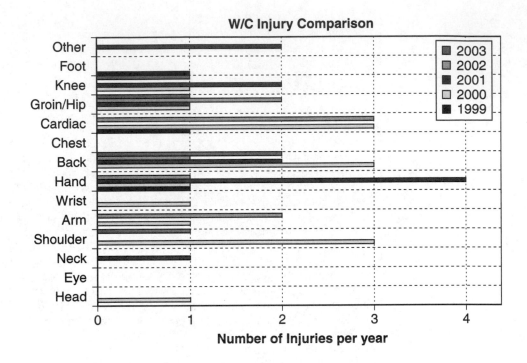

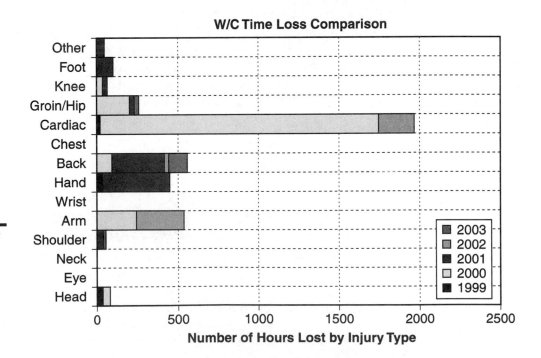

Figure 12–4 *Annual activity reports should include safety and health information.* (Courtesy Palm Harbor Fire Rescue.)

program. The annual health and safety report should contain the following information, keeping in mind, again, the confidentiality of medical records:

- Introduction
- General state of the organization in terms of safety and health
- Accomplishments/improvements/benchmarks of the safety and health program for the reporting time period
- Goals and objectives for the next reporting period
- The analysis of the injury and fatalities for the reporting years
- A comparison to the national experience
- A comparison to similar departments in similar geographic areas
- A report on significant incidents, including findings and changes made for preventing future occurrences
- A summary of the organization's compliance with regulations and standards
- Other plans for improvement and resources requested
- A summary
- Specific graphics relating to the presentation of the information

Sometimes the health and safety officer will be called on to officially present this report to the senior management or the elected officials. In such cases, the presenter should have adequate audiovisual aids to assist in this process, which might include slides, overheads, or computer-generated graphics, charts, bulleted goal and objective lists, and sometimes pictures.

ACCESSING HEALTH AND SAFETY INFORMATION USING THE INTERNET

■ Note

The Internet, a network that interconnects computers worldwide, can provide access to other organizations' information, access to state and federal regulations, and access to annual national reports, such as those available through the USFA.

We are living in the information age. With electronic mail, fax machines, and video conferencing, communication is more and more becoming instantaneous. Access to information continues to grow. The safety and health program manager, through the use of a personal computer, can have access to a great deal of information to assist with the safety and health program. Through the use of an organization's e-mail system, accident and injury reports can be filed more timely, and policy changes can be distributed to all work locations with the touch of a computer key. Granted not all organizations have access to some of these systems, but for those that do, the amount and the quality of the information is invaluable.

The Internet, very simply, a network that interconnects computers world wide, can provide access to other organizations' information, access to state and federal regulations, and access to annual national reports such as those available through the USFA.

In order to access the Internet, an organization must have an Internet provider. The provider normally provides access and, in return, charges a monthly

search engines
programs on the Internet that allow a user to search the entire Internet for key words or phrases

hits
number of documents found when searching for a key word or phrase using a search engine

links
used on Internet pages so that each page may be directly tied to a page on another Internet site

fee. Once the Internet has been accessed, **search engines** can be used to search for documents or web pages that would have the words or phrases being searched for. For example, suppose a safety program manager wanted to obtain more information about exposures to bloodborne pathogens. Using the search feature of the Internet and typing in the words *bloodborne* and *pathogens.* the search engine searches the Internet for documents and web pages containing these words. The search engine then displays a list of **hits**. The user can then click on any of the documents on the list and the Internet will connect them to that site.

In the following list, several popular sites for safety and health information are described. Each also commonly has **links** to other pages with similar information. The reader is encouraged to visit these sites and to search for others. The information gained is timely and useful to the safety program. However, a note of caution: Information on the Internet is not guaranteed to be factual. If there is any doubt, the user should verify the validity of the information. Often the search produces so many hits that it would be unreasonable to look at all of them. In this case, the user may want to be more specific. In our example, the next search may have the words bloodborne, pathogen, and exposure.

Some examples, or common sites of interest to the safety and health manager are described in Text Box 12–2.

A book could be written on the Internet and its relation to emergency services. Instead, Appendix B lists sites pertinent to the health and safety of emergency responders. This list is not exhaustive but a good start.

Text Box 12–2   *Some example sites that should be visited by the safety and health manager.*

- The Department of Labor has a Web site that gives access to OSHA regulations and a history of organizations that were cited for specific violations. http://www.osha.gov

- The International Association of Fire Chiefs and the International Association of Fire Fighters each have Web sites that can provide usable information. http://www.iafc.org and www.iaff.org

- FEMA, the USFA, and the National Fire Academy have a very comprehensive site with a great deal of health and safety information and where to find resource information. http://www.usfa.fema.gov

- The Centers for Disease Control and Prevention site can be helpful in obtaining information regarding infectious diseases and information on how to comply with regulations dealing with infectious diseases. This site is also where NIOSH is hosted. http://www.cdc.gov

- The NFPA provides many of the safety and health standards. http://www.nfpa.org

- There are numerous sites for fire/EMS magazines and journals.

- Some worker's compensation carriers are allowing for Internet access to an organization's claim records. The access is limited to those who need the information for tracking claims or who are responsible for the health and safety program.

SUMMARY

Data collection and information analysis are important components of the safety and health program. The data collection needs will vary based on whether the information is for use within the department or for an external agency. Each can be turned into valuable feedback that will assist in the evaluation of the program. Regardless of where the information is to be used, a form or similar consistent means should be established for gathering the data. On scene, information can begin to accumulate.

Once analyzed, the information should be compiled into an annual report for distribution organizationwide. The annual report has a number of required sections and should answer the questions of where were we, where are we now, where are we going, and how can we get there?

By using computers and the Internet, an organization can communicate much faster and can access much more information more easily than has ever been possible. Many national, state, and local health and safety-related agencies have information on the Internet, as do local emergency response organizations. This information can be a valuable tool in assisting the health and safety program manager and safety committee in program design, training, and evaluation.

Concluding Thought: As with the other components of the program, information management is essential. Through the use of computers and the Internet, there is virtually no limit to the access of this information and new information. ICS and Osha logs are good starting points.

REVIEW QUESTIONS

1. List three reasons for data collection and reporting.
2. Compare data collection from the internal perspective to the external perspective.
3. Injury data downloaded from the USFA Web site would be considered internal data.
   A. True
   B. False
4. Which of the following are essential to an organization's information system?

A. Data must be collected using a common means.
B. Data must be stored so that it is easily retrievable.
C. Data must be analyzed and put in a useful format.
D. All of the above are essential components.

5. List the sections that should be a part of the annual safety and health report.

ACTIVITIES

1. Compare the annual health and safety report from your department with the recommendations in this text. Are all the sections included? Is the document used as a planning tool?
2. Analyze the procedures in your department for collecting and storing health and safety data. Are the data retrievable in a usable format? Are the data useful as information?
3. Access the Internet if possible. Visit the sites that are recommended in the text and search for others. What information can be useful in your department?

Chapter 13

SPECIAL TOPICS IN EMERGENCY SERVICE OCCUPATIONAL SAFETY AND HEALTH

Learning Objectives

Upon completion of this chapter, you should be able to:

■ Describe the legal considerations of the safety and health program.

■ Describe ethical considerations of the safety and health program.

■ Describe financial considerations of the safety and health program.

■ Describe health and safety considerations in the diversified workforce.

■ Describe the safety and health program implications of future trends and technologies.

CASE REVIEW

On September 25, 2001, a 19-year-old volunteer fire-fighter died and two volunteer firefighters were injured during a multiagency, live-burn training session. The firefighter who died and one of the firefighters that was burned were playing the role of firefighters who had become trapped on the second level of the structure. The training became reality when the fire was started and progressed up the stairwell, accelerated by a foam mattress that was ignited on the first floor. The victim and one injured firefighter were recovered from the second-level front bedroom where they had been placed for the training. The other firefighter who was burned jumped from a second-level window in the rear bedroom. The firefighter who died was unresponsive when removed from the structure. Advanced life-saving procedures were initiated en route to the local hospital where he was pronounced dead. The two injured firefighters suffered severe burns and were air-lifted to an area burn unit. NIOSH investigated this incident and concluded that to minimize the risk of similar occurrences, fire departments should ensure the following:

- No one plays the role of victim inside the structure during live-burn training

- A certified instructor is in charge of the live-burn training

- A separate safety officer is appointed and has the authority to intervene and control any aspect of the operation

- Only one training fire is lit at a time by a designated ignition officer and a charged hoseline is present while igniting the fire

- Standard operating procedures are developed and followed

- All firefighters participating in live-burn training have achieved a minimum level of basic training

- Before conducting live-burn training, a preburn briefing session is conducted and an evacuation plan and signal are established for all participants

- Fires used for live-burn training are not located in any designated exit paths

- The fuels used in the live-burn training evolutions have known burning characteristics and the structure is inspected for possible environmental hazards

- States should develop a permitting procedure for live-burn training to be conducted at acquired structures and should ensure that all the requirements of *NFPA 1403* have been met before issuing the permit.

After this incident, the assistant chief of the fire department was charged with second-degree manslaughter. The case was both emotionally charged and received national attention from the fire service. In the end, the assistant chief was found guilty of criminally negligent homicide and was sentenced to 75 days in jail and required to avoid contact with any fire department under a 5-year term of probation.

INTRODUCTION

This text and study of occupational safety and health would not be complete without a discussion of legal, ethical, and financial issues that could affect the safety and health program. This chapter discusses these issues and how an increasingly diverse workforce and future trends and technologies may affect the safety and health program.

Some of the future trends presented are forward looking and therefore should not be considered complete. Instead, they are designed to spark interest, future research, and to produce changes in the emergency service field in order to be prepared for the challenges that are sure to be ahead.

LEGAL ISSUES

A number of legal issues affect emergency service organizations. Legal action against emergency service organizations were once uncommon; however, there has been a significant increase in these cases. Some of these legal issues can have a direct impact on or relationship to the safety and health program. For example, the safety program requirement for emergency vehicle operators could be called into question after an apparatus is involved in an accident. The safety and health program manager should be aware of some of the more common issues related to the safety and health program. The following snapshot of some of these issues should not be considered all-inclusive, nor should it be considered legal advice, it is simply meant to provide awareness. When specific issues arise, the department's legal counsel must be contacted.

Tort Liability

The concept of liability is important to the safety and health manager as well as department members. The safety and health program could be questioned should a claim of liability arise. **Tort liability** is an action that harms another person or group and can occur when a person or group of people act, or fail to act, without right, and thus harm another directly, or indirectly. The following four types of tort liabilities are applicable to the safety and health program.

tort liability

an action that harms another person or group and can occur when a person or group of people act, or fail to act, without right, and thus harm another directly, or indirectly

- *Strict Liability.* Violation of the law or other regulations, even if violation is unintentional
- *Intentional Liability.* Violation of the law or other regulations knowingly and harm results
- *Negligent Liability.* Person fails to do what a reasonable and prudent person would have done under the same or similar circumstances
- *Warrant Liability.* Promised service level is not delivered and harm results

It is easy to see where these types of liabilities can have an impact on the safety and health program. For example, suppose a department is required to follow OSHA 1910.134, Respiratory Protection Regulation, of which SCBA mask fit testing is a component. If the department does not fit test its employees, is it guilt of strict or intentional liability? Maybe. In order to prove liability, four elements must be present:

- Duty or standard to act
- Breach of duty, either an action or omission

- Failure caused the harm
- Actual measurable loss or harm

In the SCBA example, if an employee had been injured as a result of smoke inhalation because of a poor fitting mask, would the department have liability?
Text Box 13–1 gives some examples of ways that liability can arise.

Text Box 13–1   *Situations that can result in a liability claim.*

- Poor or out-of-date plans
- Personnel not trained; undocumented or unsafe training
- Hazards and risk not identified
- Personnel not warned of hazards
- Equipment not operated or maintained properly
- Standards not present or enforced
- Failure to follow the standard of care (refer to Chapter 2)

■ **Note**

The emergency service administration and safety and health officer must have a close working relationship with the legal staff.

The issue of liability is complex and changing with every court decision. The emergency service administration and safety and health officer must have a close working relationship with the legal staff.

American with Disabilities Act

Title I of the Americans with Disabilities Act (ADA) of 1990 took effect July 26, 1992. This act prohibits private employers, state and local governments, employment agencies, and labor unions from discriminating against qualified individuals with disabilities in job application procedures, hiring, firing, advancement, compensation, job training, and other terms, conditions, and privileges of employment. An individual with a disability is a person who has one or more of the following:

■ **Note**

An organization may not ask job applicants about the existence, nature, or severity of a disability.

- Has a physical or mental impairment that substantially limits one or more major life activities
- Has a record of such an impairment
- Is regarded as having such an impairment

■ **Note**

Medical examinations of employees must be job related and consistent with the employer's business needs.

The emergency service safety and health manager should understand the provisions in Title I of the ADA. One important provision is that of medical examinations and disability. Specifically, an organization may not ask job applicants about the existence, nature, or severity of a disability. Applicants may be asked about their ability to perform specific job functions. A job offer may be conditioned on the results of a medical examination, but only if the examination is required for all entering employees in similar jobs. Medical examinations of employees must be job related and consistent with the employer's business needs. Again, it is necessary that the safety and health program comply with this law.

Age Discrimination in Employment Act

The Age Discrimination in Employment Act of 1967 (ADEA) is another federal regulation that should be a consideration in the safety and health program. The ADEA protects individuals who are 40 years of age or older from employment discrimination based on age and applies to both employees and job applicants. Under the ADEA, it is unlawful to discriminate against a person because of age with respect to any term, condition, or privilege of employment including, but not limited to, hiring, firing, promotion, layoff, compensation, benefits, job assignments, and training. The ADEA applies to employers with 20 or more employees, including state and local governments. It also applies to employment agencies, labor organizations, and the federal government.

ETHICAL ISSUES

Ethics could be defined simply as doing the right thing. There are ethical considerations in every business or organization. There are also ethical issues for the safety and health manager and others involved in the safety and health program. One safety and health ethical concept is that members of an organization should be provided with the safest work environment possible. In emergency services, in terms of incident response and operation, we seldom have control of the work environment; however, in terms of operation, facilities, and equipment, the organization can do everything possible to ensure the members' safety.

Simply in the course of accident investigation and record management, the members involved directly with the program, including the safety and health officer, management, and the safety committee, are often privy to confidential information. Members of the program that have access to this information must know that specifics of events and other information about coworkers should not be discussed with other members. The trust and respect of confidential information is essential for the success of the program. Members involved in the safety and health program must follow the ethical standards required by the organization; however, they must understand and be educated in further ethical standards that can be applied to them as a result of their relationship to the safety and health program.

FINANCIAL ISSUES

As described in the preceding chapters, components of safety and health programs cost money; however, the cost of not having a comprehensive program is more than the costs associated with program operations. There are some common methods for funding the program. The first and most common is to use funds from the organization's annual budget. Another is to seek grant funding. Often to get the necessary funds appropriated for a safety and health program, either through the annual budget or through grant funding, a cost-benefit analysis must be performed and presented.

Annual Budget Process

The most common funding source is an organization's annual budget. The safety and health officer, as part of the senior staff, should have input into the budget preparation. The safety and health officer is often in a position of salesperson, selling the benefits of a particular part of the program for which they are seeking funding. A complete cost-benefit analysis is necessary and helpful to persuade funders of the necessity for appropriation of funds.

Aside from new components that the safety and health officer may propose each year, ongoing costs must be considered, including:

- Salary and benefits for the safety and health officer
- Annual medical exams for personnel
- Fitness equipment or program maintenance
- Funds for safety and health training of personnel
- Payment for medical services and department physician
- PPE replacement and maintenance

The safety and health officer has to anticipate and project these costs each year. Most are operating costs from the budget, but there will be capital items as well, such as fitness equipment. There are some opportunities for grant or foundation funding for starting components of the safety and health program, but this grant funding is often limited to capital items.

Grant Funding

There are a number of funding sources from grants. Some are federal, some state, and even some at the local city or county level. Sources of grant funding can be researched on the Internet: http://www.grants.gov is one such site. There are also subscriptions to grant newsletters and even consultants that can help a department look for funds and write the grant application, but consultants normally charge money for their services. Federal grant programs are published in the *Federal Register*. Safety and health managers should focus on the proposed project and look at emergency service-based grants as well as those outside of the emergency field. A department received a technology grant for laptop computers and a GIS mapping and preplanning system for its apparatus. This enhanced information was available at the emergency scene, hence it helped enhance safety.

■ **Note**
Often to get the necessary funds appropriated for a safety and health program, either through the annual budget or through grant funding, a cost-benefit analysis must be performed and presented.

■ **Note**
The most common funding source is an organization's annual budget.

■ **Note**
There are a number of funding sources from grants. Some are federal, some state, and even some at the local city or county level.

Often the grant applicant must show how funds will be allocated to continue the program after the grant period ends. Some grants require a local match, some do not. Some examples of emergency services-related grant programs are listed in Text Box 13–2. This is an brief example list only and should not be considered all-inclusive.

Text Box 13–2   *Common emergency service-related grant programs.*

- *Assistance to Firefighters Grant Program (AFGP).* Administered by the Department of Homeland Security's Office for Domestic Preparedness, this program assists rural, urban, and suburban fire departments throughout the United States. These funds are used by the nation's firefighters to increase the effectiveness of firefighting operations, to improve firefighter health and safety programs, and to establish or expand fire prevention and safety programs. This program was know as the Fire Investment and Response Enhancement (FIRE) Act and was moved from FEMA to the Department of Homeland Security in 2004.

- *Department of Justice.* Has grants available for terrorism preparedness equipment and training.

- *Volunteer Fire Assistance (VFA).* The purpose of this program, formerly known as the Rural Community Fire Protection (RCFP) Program, is to provide federal financial, technical, and other assistance to state foresters and other appropriate officials to organize, train, and equip fire departments in rural areas and rural communities to prevent and suppress fires. A rural community is defined as having a population of 10,000 or less.

- *Hazardous Materials Emergency Preparedness (HMEP).* This grant program is intended to provide financial and technical assistance as well as national direction and guidance to enhance state, territorial, tribal, and local hazardous materials emergency planning and training. The HMEP grant program distributes fees collected from shippers and carriers of hazardous materials to emergency responders for hazmat training and to local emergency planning committees (LEPCs) for hazmat planning.

SAFETY CONSIDERATIONS IN THE DIVERSIFIED WORKFORCE

According to the report *Workforce 2000*,[1] the diversity in the workplace will continue to increase. Although this document was written for the workforce as a whole, there are implications for the emergency service occupations. Some of the findings from the report include:

- Population growth will slow.
- The average age of the workforce will rise.

---

[1]Published in 1987 by the Hudson Institute of Indianapolis for the Department of Labor

- More women will enter the workforce.
- White males will be a smaller percentage of the workforce.
- Minorities will be a larger share of the new entrants into the labor force.

Although there are not safety and health considerations for all of these findings, a couple do support some discussion. The increasing average age of the workforce will require continued emphasis on physical fitness. Physical abilities testing should be based on job-validated requirements. The IAFC/IAFF joint fitness and wellness initiative group has developed one such abilities test, the Candidate Physical Abilities Test or CPAT. The CPAT validated abilities test consists of a timed course evaluating performance on eight events. Information on the CPAT is available from the IAFC or through IAFF locals.

Although the emergency services have generally adapted well to women in the workforce, their entry has been relatively recent. What safety and health considerations must be undertaken for women who are pregnant? Policies must be developed before the situation arises. Many chemicals that can affect the unborn fetus as well as the mother and heavy lifting may also be a concern.

The design and location of equipment on apparatus must be considered. Part of a diverse workforce is variations in height. Years ago, fire departments had height requirements. After these were challenged and removed, there became a need to redesign apparatus, particularly ladder storage, patient compartment height, and the height of hose beds.

Equipment purchase recommendations should consider the diversity in personnel. OSHA requires fit testing of SCBA masks. It is unlikely that one size of masks on the apparatus would provide a proper fit for all members. The same holds true for other PPE; the organization must make available a variety of sizes of PPE, as protective clothing must fit well and comfortably without causing undue heat stress or adding excessive weight. A firefighter with improper fitting protective equipment will be inefficient and may be more prone to injury.

These are but a few examples of safety and health considerations in a diversified workforce. The safety and health manager must maintain awareness of trends in these areas and be able to adjust the safety and health program as needed.

FUTURE TRENDS AND TECHNOLOGIES

Probably one of the more exciting prospects in the field of safety and health is looking toward the future. Just consider for a moment the technological changes that have occurred over the past 20 years. Look at the protective equipment changes and how the emergency services have had to adapt. Many items are being manufactured now that are somewhat expensive and out of reach of some departments, but as technology improves, the market will drive the price down. Remember the first personal computers? Compare their price and capabilities to what you can buy today for a fraction of the cost. Ten years ago, few departments had a thermal imaging camera. Now that costs have come down some and additional grant funding has become available, many departments have one camera on each apparatus. Some issues that could impact the safety and health program in the future are given in the following list:

- Emergency Scene Geographic Information System (GIS) interface with the GPS linked to the incident commander's computer for location of all personnel
- Recruitment and retention of qualified personnel
- Terrorism preparedness causing a greater need to work closely with other related response agencies and creating the need for interoperability of communications equipment and better coordination of IMS
- Individual status information projected on the SCBA mask and wirelessly to command officers
- The ability to have building floor plans projected on the SCBA facemask
- Better early warning for fires and better private fire protection
- Greater use of GPS and intersection control to reduce intersection accidents
- Even greater use of thermal imaging systems
- Further improvements in protective clothing to prevent exposures
- Change from compressed air to cryogenic air supply, reducing the weight and increasing the duration of SCBA
- Greater emphasis on occupational safety. Hiring of specific health and safety officers with higher-level degrees in occupational safety and health.
- A national interest in emergency service safety and health
- More emergency service safety and health-related education and training
- A greater emphasis on research for health and safety programs including a study of what is done in other countries, giving the safety and health problem a global aspect

SUMMARY

The safety and health concepts presented in this book would not be complete without a look at legal, ethical, and financial issues. The legal issues are best dealt with using legal counsel; however, the safety and health program manager should have a working knowledge of some legal concepts, specifically those involving negligence and compliance with applicable federal regulations. Ethics are a part of every organization. Members active with the safety and health program, as a result of their position, may be held to a higher ethical standard. The financial issues involve funding for recurring annual costs and costs associated with program component startup. The safety and health manager should have input into the budget process. The annual budget is the most common funding source, but alternative funding, such as grants do exist.

Changes in the diversity and demographics will have implications to the safety and health program manager and to the overall program. These changes have been identified in the *Workforce 2000* document and focus on a slowing population growth, a rise in the average age on the workforce, more woman entering the workforce, white males becoming a smaller percentage of the workforce, and minorities becoming a larger share of the new entrants into the workforce. There are health and safety considerations in a couple of these areas.

Future technology and trends will affect the safety and health program. From better protective equipment to the safety and health manager as a professional similar to a fire protection engineer, the focus will be on continued support and improvement of the safety and health for the emergency responders of the world.

Concluding Thought: There are a number of special considerations in the development and management of the safety and health program. The safety and health officer must have a working knowledge in a variety of subjects broader than just emergency service operations.

REVIEW QUESTIONS

1. List the predictions of *Workforce 2000* and identify those that will have a health and safety impact.

2. Describe two funding sources for the safety and health program.

3. List five implications of future technology of a safety and health program.

4. Describe the four kinds of tort liability presented in this textbook.

5. What are the implications of an older workforce on the health and safety program?

ACTIVITIES

1. Examine the *Workforce 2000* predictions. How will they affect your department?

2. Review any grant funds that your department has received. If none, what opportunities do you see?

3. Pick any two future technologies or trends and discuss their implications for your department.

REPRINT OF THE 2002 FIREFIGHTER FATALITY SURVEY FROM THE UNITED STATES FIRE ADMINISTRATION

Although this text only includes the report from 2002, these reports are done annually. The report is included here to introduce the reader to the information available from a report of this nature. The reader is encouraged to obtain a current report from the publications center of the USFA or from http://www.usfa.fema.gov for local comparison. The complete report includes an appendix that is a summary of individual incidents. This summary is not included here. (Courtesy of Federal Emergency Management Agency (FEMA).)

Firefighter Fatalities
in the United States
in 2002

U.S. Department of Homeland Security
Federal Emergency Management Agency
U.S. Fire Administration

July 2003

In memory of all firefighters who
answered their last call in 2002

To their families and friends

To their service and sacrifice

Courtesy of Federal Emergency Management Agency (FEMA)

U.S. Fire Administration Mission Statement

As an entity of the Federal Emergency Management Agency, the mission of the United States Fire Administration is to reduce life and economic losses due to fire and related emergencies, through leadership, advocacy, coordination, and support. We serve the Nation independently, in coordination with other Federal agencies, and in partnership with fire protection and emergency service communities. With a commitment to excellence, we provide public education, training, technology, and data initiatives.

On March 1, 2003, FEMA became part of the U.S. Department of Homeland Security. FEMA's continuing mission within the new department is to lead the effort to prepare the nation for all hazards and effectively manage federal response and recovery efforts following any national incident. FEMA also initiates proactive mitigation activities, trains first responders, and manages Citizen Corps, the National Flood Insurance Program and the U.S. Fire Administration.

ACKNOWLEDGEMENTS

This study of firefighter fatalities would not have been possible without the cooperation and assistance of many members of the fire service across the United States. Members of individual fire departments, chief fire officers, the National Interagency Fire Center, United States Forest Service personnel, the United States military, the Department of Justice, NFPA International, and many others contributed important information for this report.

IOCAD Emergency Services Group of Emmitsburg, Maryland (a division of IOCAD Engineering Services, Inc.) conducted this analysis for the United States Fire Administration (USFA) under contract EME-98-CO-0202-T017.

The ultimate objective of this effort is to reduce the number of firefighter deaths through an increased awareness and understanding of their causes and how they can be prevented. Firefighting, rescue, and other types of emergency operations are essential activities in an inherently dangerous profession, and unfortunate tragedies do occur. This is the risk all firefighters accept every time they respond to an emergency incident. However, the risk can be greatly reduced through efforts to increase firefighter health and safety.

Photographic Acknowledgments

The USFA would like to extend its thanks to the following individuals for providing photographs for this report:

John Severson, Indianapolis Star

Firefighters lead a procession for Indianapolis Private Paul Kelly Jolliff as it enters the cemetery III

Patrick Schneider, Charlotte Observer

Firefighter Joshua Early was killed when the first floor of this structure collapsed and he fell into the fire-involved basement 12

J. Intintoli, Daily News Journal

This photo depicts the aftermath of a tanker crash that killed Firefighter Jason Kevin Jackson on September 5, 2002 14

ACKNOWLEDGEMENTS

ACKNOWLEDGEMENTS

TABLE OF CONTENTS

TABLE OF CONTENTS

TABLE OF CONTENTS

TABLE OF CONTENTS

BACKGROUND

For 26 years, the United States Fire Administration (USFA) has tracked the number of firefighter fatalities and conducted an annual analysis. Through the collection of information on the causes of firefighter deaths, the USFA is able to focus on specific problems and direct efforts toward finding solutions to reduce the number of firefighter fatalities in the future. This information is also used to measure the effectiveness of current programs directed toward firefighter health and safety.

One of the USFA's main program goals is a 25-percent reduction in firefighter fatalities in 5 years and a 50-percent reduction within 10 years. The emphasis placed on these goals by the USFA is underscored by the fact that these goals represent one of the five major objectives that guide the actions of the USFA.

In addition to the analysis, the USFA provides a list of firefighter fatalities to the National Fallen Firefighters Foundation. If Memorial criteria are met, the fallen firefighter's next of kin, as well as members of the individual fire department, are invited to the annual Fallen Firefighters Memorial Service. The service is normally held at the National Emergency Training Center (NETC) in Emmitsburg, Maryland, during Fire Prevention Week. Additional information regarding the Memorial Service can be found on the Internet at http://www.firehero.org/ or by calling the National Fallen Firefighters Foundation at (301) 447-1365.

Other resources and information regarding firefighter fatalities, including current fatality notices, the National Fallen Firefighters Memorial database, and links to the Public Safety Officer Benefit (PSOB) program can be found at http://www.usfa.fema.gov/inside-usfa/ffmem.cfm.

INTRODUCTION

This report continues a series of annual studies by the USFA of onduty firefighter fatalities in the United States.

The specific objective of this study is to identify all onduty firefighter fatalities that occurred in the United States in 2002 and to analyze the circumstances surrounding each occurrence. The study is intended to help identify approaches that could reduce the number of firefighter deaths in future years.

In addition to the 2002 overall findings, this study includes a study of firefighters killed while responding in their personal vehicles and low cost steps that can be taken to prevent the loss of firefighter lives.

Who Is a Firefighter?

For the purpose of this study, the term firefighter covers all members of organized fire departments in all States, the District of Columbia, and the Territories of Puerto Rico, the Virgin Islands, American Samoa, the Commonwealth of the Northern Mariana Islands, and Guam. It includes career and volunteer firefighters; full-time public safety officers acting as firefighters; State, Territory, and Federal government fire service personnel, including wildland firefighters; and privately employed firefighters, including employees of contract fire departments and trained members of industrial fire brigades, whether full- or part-time. It also includes contract personnel working as firefighters or assigned to work in direct support of fire service organizations.

Under this definition, the study includes not only local and municipal firefighters but also seasonal and full-time employees of the United States Forest Service, the Bureau of Land Management, the Bureau of Indian Affairs, the Bureau of Fish and Wildlife, the National Park Service, and State wildland agencies. The definition also includes prison inmates serving on firefighting crews; firefighters employed by other governmental agencies, such as the United States Department of Energy; military personnel performing assigned fire suppression activities; and civilian firefighters working at military installations.

What Constitutes an Onduty Fatality?

Onduty fatalities include any injury or illness sustained while on duty that proves fatal. The term "on duty" refers to being involved in operations at the scene of an emergency, whether it is a fire or nonfire incident; responding to or returning from an incident; performing other officially assigned duties such as training, maintenance, public education, inspection, investigations, court testimony, and fundraising; and being on-call, under orders, or on standby duty except at the individual's home or place of business. An individual who experiences a heart attack or other fatal injury at home as he or she prepares to respond to an emergency is considered on duty when the response begins. A firefighter that becomes ill while performing fire department duties and suffers a heart attack shortly after arriving home or at another location may be considered on duty since the inception of the heart attack occurred while the firefighter was on duty.

A fatality may be caused directly by an accidental or intentional injury in either emergency or nonemergency circumstances, or it may be attributed to an occupationally-related fatal illness. A common example of a fatal illness incurred on duty is a heart attack. Fatalities attributed to occupational illnesses would also include a communicable disease contracted while on duty that proved fatal when the disease could be attributed to a documented occupational exposure.

Injuries and illnesses are included even when death is considerably delayed after the original incident. When the incident and the death occur in different years, the analysis counts the fatality as having occurred in the year in which the incident took place.

One firefighter died in 2002 as the result of injuries he suffered in 1982. The USFA was notified of the deaths of two firefighters in 2001 that were not known or included in the firefighter fatality report for that year. One of the firefighters who died in 2001 was injured at a fire in 1997. For statistical purposes, each firefighter death is counted in the year in which the incident occurred. Information about these three deaths is included in the appendix of this report, but they are not addressed in the body of the report unless the death impacts retrospective statistical comparisons.

There is no established mechanism for identifying fatalities that result from illnesses such as cancer that develop over long periods of time, which may be related to occupational exposure to hazardous materials or products of combustion. It has proved to be very difficult over the years to provide a complete evaluation of an occupational illness as a causal factor in firefighter deaths due to the following limitations: the exposure of firefighters to toxic hazards is not sufficiently tracked, the often delayed long-term effects of such toxic hazard exposures, and the exposures firefighters may receive while off duty.

Sources of Initial Notification

As an integral part of its ongoing program to collect and analyze fire data, USFA solicits information on firefighter fatalities directly from the fire service and from a wide range of other sources. These sources include the Public Safety Officers' Benefit (PSOB) program administered by the Department of Justice, the National Institute for Occupational Safety and Health (NIOSH), the Occupational Safety and Health Administration (OSHA), the United States military, the National Interagency Fire Center, and other Federal agencies.

The USFA receives notification of some deaths directly from fire departments, as well as from such fire service organizations as the International Association of Fire Chiefs (IAFC), the International Association of Fire Fighters (IAFF), NFPA International, the National Volunteer Fire Council (NVFC), State fire marshals, State training organizations, other State and local organizations, fire service Internet sites, news services, and fire service publications. The USFA also keeps track of fatal fire incidents as part of its Major Fires Investigation Program and performs an ongoing analysis of data from the National Fire Incident Reporting System (NFIRS).

INTRODUCTION

Procedure for Including a Fatality in the Study

In most cases, after notification of a fatal incident, initial telephone contact is made with local authorities by the USFA to verify the incident, its location, jurisdiction, and the fire department or agency involved. Further information about the deceased firefighter and the incident may be obtained from the chief of the fire department or his or her designee over the phone or by other data collection forms.

Information that is requested routinely includes NFIRS-1 (incident) and NFIRS-3 (fire service casualty) reports, the fire department's own incident reports and internal investigation reports, copies of death certificates or autopsy results, special investigative reports, police reports, photographs and diagrams, and newspaper or media accounts of the incident. Information on the incident may also be gathered from NFPA International, the USFA, or NIOSH reports on an incident.

After obtaining this information, a determination is made as to whether the death qualifies as an on duty firefighter fatality according to the previously described criteria. The same criteria were used for this study as in previous annual studies. Additional information may be requested, either by follow-up with the fire department directly, from State vital records offices, or other agencies. The determination as to whether a fatality qualifies as an on duty death for inclusion in this statistical analysis is made by the USFA. The final determination as to whether a fatality qualifies as a line-of-duty death for inclusion in the Fallen Firefighters Memorial Service is made by the National Fallen Firefighters Foundation.

INTRODUCTION

2002 FINDINGS

One hundred firefighters died while on duty in 2002. While the number of deaths in 2002 is dramatically lower than the horrendous loss of firefighter lives in 2001, it is still an unacceptable level of loss. The level of firefighter fatalities trended downward in the early 1990s but has settled at somewhat higher levels in the latter part of the 1990s and into this century.

Figure 1. Onduty Firefighter Fatalities (1977-2002)

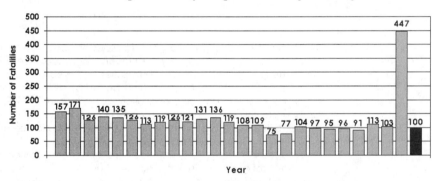

With the deaths of 100 firefighters in 2002, this is the fifth time in the last 10 years and the tenth time within the last 15 years when the total number of firefighter fatalities has reached or exceeded 100. The lowest years on record are 1992 with 75 fatalities and 1993 with 77 fatalities (Figure 1).

The 100 deaths resulted from a total of 84 incidents. There were nine multiple firefighter fatality incidents.

In 2001, 344 firefighters were killed as a result of the attacks on the World Trade Center (WTC) in New York City on September 11. When conducting multi-year comparisons of firefighter fatalities in this report, it may be necessary to set these deaths apart for illustrative purposes. This action is by no means a minimization of the supreme sacrifice made by these firefighters.

Seven Firefighters were murdered in 2002:

Six in arson-caused or suspicious fires

One at the hands of a gunman who had set a structure fire.

The median age for firefighters who died while on duty in 2002 was 41 years and 9 months.

Figure 2. Firefighter Fatalities per 100,000 Fire Incidents

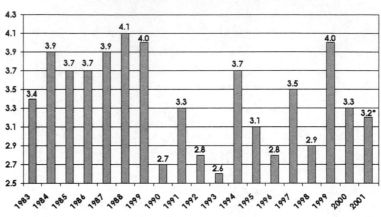

*Does not include WTC

While the total number of firefighter fatalities has been decreasing over the past 20 years, the number of firefighter deaths per fire incident has actually risen. The chart above (Figure 2) compares the total number of firefighter fatalities each year that are associated with responses to fires and the total number of fire incidents reported by NFPA International through 2001 (2002 data is not yet available). Despite a downward dip in the early 1990s, the level of firefighter fatalities is back up to the same levels experienced in the 1980s. If the firefighter deaths at the WTC are included in the 2001 data, the number rises to 23.1 firefighter fatalities per 100,000 fires.

A retrospective study of firefighter fatalities reporting on firefighter fatalities over a 10-year period entitled "Firefighter Fatality Retrospective Study 1990-2000" is available at no cost from the USFA. The report may also be found on the internet at www.usfa.fema.gov/inside-usfa/fa-220.cfm.

Career and Volunteer Fatalities

Firefighter fatalities in 2002 includes 66 volunteer firefighters and 34 career firefighters (Table 1). Among the volunteer firefighter fatalities, 50 were from local or municipal volunteer fire departments, and 16 were seasonal or contract members of wildland fire agencies. All of the career firefighters that died were members of local or municipal fire departments. Ninety-five of the fatalities were men and 5 were women.

Table 1. Career vs. Volunteer Fatalities

TOTAL (443)	
Career (34)	**Volunteer (66)**
Metro Depts (18)	Suburban/Urban VFD (22)
Other Depts (16)	Rural VFD (28)
	Wildland Seasonal/Part-time (16)

2002 FINDINGS

Multiple Firefighter Fatality Incidents

The 100 fatalities resulted from 84 incidents. There were 9 multiple firefighter fatality incidents resulting in the deaths of 25 firefighters (Table 2).

In 2002, five firefighters died when the van in which they were riding was involved in a crash as it headed to join the firefighting effort in Colorado; three firefighters died in the crash of a wildland firefighting aircraft in California; three firefighters died in a structural collapse as they searched a building for victims in New Jersey; three California firefighters died when their wildland engine apparatus left the roadway and rolled down a steep embankment; three Oregon firefighters died as they fought a fire in a commercial building; two New York firefighters died as they advanced a hoseline into a fire-involved structure -- the floor collapsed and they fell into the fire area; two firefighters died during a structural firefight in Missouri; two wildland firefighters died in the crash of their airtanker in Colorado; and two Florida firefighters died during structural firefighting training.

Table 2. Multiple Fatality Incidents

Year	Number of Incidents	Total Number of Fatalities
2002	9	25
2001	8	362
2001 w/o WTC	7	18
2000	5	10
1999	6	22
1998	10	22
1997	8	17
1996	3	8

Wildland Firefighting Fatalities

The number of deaths associated with brush, grass, or wildland firefighting in 2002 was 23 (Table 3). In 2002, there were six firefighter deaths associated with wildland aircraft firefighting duties (two multiple fatality incidents claimed five firefighters and one crash claimed a single firefighter). This total includes fixed-wing aircraft and helicopters (Table 4).

> *In 2002, there were no firefighter fatalities as the result of firefighters having their positions overrun by wildland fires.*

Five firefighters based in Oregon died in the crash of a passenger van in Colorado; three California firefighters died when their engine left the roadway and rolled; two firefighters died while completing pack tests in California and Montana; two firefighters suffered heart attacks as they fought wildland fires; two firefighters died in apparatus crashes while responding to wildland fires; two firefighters were thrown from wildland firefighting vehicles and struck; and one firefighter died when he was struck by a falling tree.

Table 3. Fatalities Associated with Wildland Firefighting

Year	Total Number of Fatalities
2002	23
2001	15
2000	19
1999	28
1998	13
1997	10
1996	5

Table 4. Wildland Firefighting Aircraft Fatalities

Year	Total Number of Fatalities
2002	6
2001	6
2000	6
1999	0
1998	3
1997	5
1996	0

2002 FINDINGS

TYPE OF DUTY

In 2002, 73 onduty firefighter deaths were associated with emergency incidents (Figure 3). This includes all firefighters who died while responding to an emergency, while at the emergency scene, or while returning from the emergency incident. Nonemergency activities accounted for 27 fatalities. Nonemergency duties include training, administrative activities, or performing other functions that are not related to an emergency incident. A 7-year historical perspective concerning the percentage of firefighter deaths that occurred during emergency duty is presented in (Table 5).

Figure 3. Firefighter Fatalities by Type of Duty (2002)

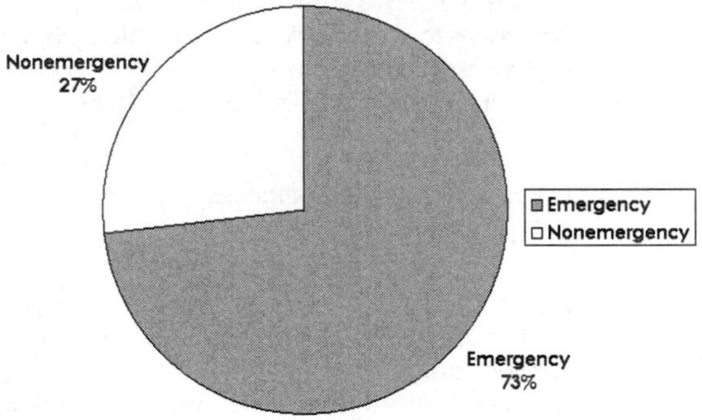

Nonemergency
27%

Emergency
Nonemergency

Emergency
73%

The number of fatalities by type of duty being performed in 2002 is shown in Table 6 and presented graphically in Figure 4. As in previous years, the largest number of deaths occurred during fireground operations. There were 45 fireground deaths, almost half of the total.

Table 5. Emergency Duty Firefighter Fatalities

Year	Percentage of All Fatalities
2002	73
2001	65
2001 w/WTC	92
2000	71
1999	87
1998	77
1997	81
1996	72

Table 6. 2002 Fatalities by Type of Duty

Type of Duty	Number of Fatalities
Fireground Operations	45
Responding/Returning from Alarm	13
Other Onduty Fatalities	14
Training	11
Nonfire Emergencies	12
After an Incident	5
Total	**100**

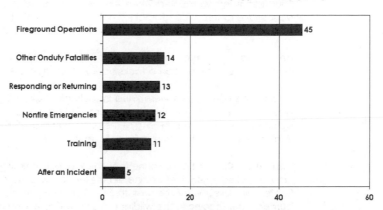

Figure 4. Fatalities by Type of Duty (2002)

Fireground Operations

Many of the firefighting deaths in 2002 were related in some way to structural collapse. Three New Jersey firefighters were killed when the building that they were searching suffered a sudden catastrophic collapse; three Oregon firefighters died when a fire-weakened roof collapsed and trapped firefighters inside of the structure; two New York firefighters were killed as they entered a structure to fight the fire and were propelled into the fire area when the floor beneath them collapsed -- a similar incident took the life of a North Carolina firefighter; one firefighter in Indiana and one firefighter in Texas were killed after being buried in the debris of falling structures; a Pennsylvania firefighter was pinned by an initial collapse (firefighters were unable to move him) and then killed in a major collapse of a home being used for storage.

TYPE OF DUTY

Four firefighters were killed when they were trapped in the interior of burning structures. Two Missouri firefighters died of smoke inhalation in a burning commercial building, and one firefighter in Pennsylvania and one firefighter in Tennessee died of smoke inhalation in residential fires.

Heart attacks claimed the lives of 12 firefighters in 2002 while they were engaged in fire-related incidents. Ten of the heart attacks occurred at structure fires and 2 occurred at wildland fires.

A total of six firefighters were killed in firefighting aircraft crashes. Three firefighters were killed when the wings of their airtanker separated from the airframe and the plane crashed; two firefighters were killed when a wing of their airtanker broke off and the aircraft crashed; and one firefighter was killed in the crash of a helicopter.

Vehicle crashes at the fire scene took the lives of five firefighters in 2002. Three California firefighters were killed when their engine slid off the roadway and rolled into a stand of trees. One firefighter was killed when he was thrown from a brush truck after a collision -- he was severely burned after landing in the wildland fire flame front; and one firefighter was killed when he was thrown from the front bumper of a wildland apparatus when the apparatus was struck by another vehicle -- the firefighter's head was crushed by the forward movement of the apparatus after the crash.

One firefighter in New Mexico was killed on the scene of an incendiary structure fire when he was shot by the person who had started the fire; a firefighter in Minnesota was killed when he was struck by a vehicle while fighting a car fire on a highway; one firefighter in Michigan was killed when he was struck by someone jumping from a fire-involved apartment building, he suffered a head injury; a Massachusetts firefighter died of respiratory failure after fighting a structure fire; a firefighter in Colorado was killed by a falling tree; and an Iowa firefighter fell through a ventilation hole into the fire area.

Other Onduty Fatalities

Fourteen deaths occurred in 2002 during other onduty activities. Five wildland firefighters were killed when the van in which they were passengers was involved in a single-vehicle crash in Colorado; one firefighter in Connecticut, one firefighter in Pennsylvania, and, one firefighter in Tennessee died after becoming ill while on-duty; two firefighters died in their sleep while on duty, both due to heart problems; one firefighter suffered a Cerebral Vascular Accident (CVA/stroke) while exercising; one firefighter died in a crash when his vehicle was struck at an intersection as he drove to a meeting; one firefighter died when the tanker she was driving back to the station after repairs left the roadway and crashed; and one firefighter was killed while lighting fireworks for a community Independence Day celebration.

TYPE OF DUTY

Responding/Returning

Thirteen firefighters died while responding to or returning from emergency incidents in 2002 (Table 7). Six firefighters died of heart attacks while in the fire station preparing to respond or while responding in their personal or fire department vehicles.

One firefighter was killed while responding to a brush fire in a pumper; she was a back seat passenger and was crushed when the apparatus was involved in a single-vehicle crash. The firefighter was wearing a seat belt.

One firefighter was killed when the tanker he was driving went off the right side of the roadway; he steered the apparatus back onto the road, but the apparatus began to skid and crashed off the right side of the road. This is an often-repeated fatal scenario for tanker fire apparatuses.

> *A firefighter that was killed in a vehicle collision while responding in 2002 had a blood alcohol level of .11, and was therefore legally intoxicated.*

> *A new USFA publication, "Safe Operations of Fire Tankers" is now available. Further information about this publication may be found on the USFA Web site at www.usfa.fema.gov/inside-usfa/vehicle.cfm.*

This photo depicts the aftermath of a tanker crash that killed Firefighter Jason Kevin Jackson on September 5, 2002.

Five firefighters died in 2002 while responding to emergencies in their personal vehicles. One firefighter failed to negotiate a turn and crashed; one firefighter's vehicle was involved in a crash when another vehicle crossed the center line of the roadway and struck the vehicle; one firefighter was killed when he was struck by a car while responding to an incident on his bicycle; two firefighters were killed while responding to the fire station in their personal vehicles. The subject of firefighter fatalities that occur while operating personally owned vehicles (POV's) will be discussed in detail in the special topics section of this report.

TYPE OF DUTY

Table 7. Fatalities While Responding to or Returning from an Incident

Year	Number of Fatalities
2002	13
2001	23
2000	19
1999	26
1998	14
1997	21
1996	22

Three of the four firefighters that died in personal motor vehicle crashes in 2002 were not wearing their seatbelts.

A USFA project is in progress that will address the safety of firefighters in all emergency vehicles, including ambulances, fire apparatus, and personally owned vehicles (POV's), as well as safety on the scene of an incident in proximity to a roadway. Completion of the project is expected in 2003. Further information about this project may be found on the USFA Web site at www.usfa.fema.gov/inside-usfa/vehicle.cfm.

Nonfire Emergencies

Twelve firefighters died while working at the scene of nonfire emergencies. This total includes three firefighters who suffered heart attacks as they provided for the safety of scenes in their roles as fire police officers; three firefighters who suffered heart attacks at Emergency Medical Services (EMS) scenes; one firefighter who became ill at an EMS scene and later died of a CVA; and one firefighter who suffered a heart attack during debris removal after a damaging storm passed through the area.

Three firefighters were struck at the scene of motor vehicle crashes. A Texas firefighter was killed by passing traffic when he arrived first on the scene of a crash and began treatment of the injured; a Florida firefighter was struck in the median of an interstate highway when a tractor-trailer truck failed to stop for the traffic backup associated with the original crash and entered the median area; and a Kansas firefighter was killed when he was struck by a fire engine as it arrived on the scene of a crash -- the engine had experienced a mechanical failure and was unable to stop.

A South Dakota firefighter was killed when he entered a molasses tank to rescue an incapacitated worker and also became incapacitated by the oxygen deficient and hydrogen sulfide contaminated atmosphere in the tank.

Training

Eleven firefighters died in 2002 while engaged in training activities (Table 8). Two Florida firefighters were killed in a structural firefighting training exercise; a Maryland firefighter died of hyperther-

TYPE OF DUTY

mia after completing physical training during recruit school in extremely high heat; an Indiana firefighter drowned during rescue diver training; an Alabama firefighter was killed when a fire truck slipped into gear and struck him; a Kentucky firefighter was killed when the tanker he was driving back from a training exercise was struck by a train; and a New York firefighter was struck by a vehicle operated by an impaired driver as the firefighter loaded hose at the completion of a training exercise -- the driver had ignored traffic control devices.

Two firefighters died in 2002 while engaged in annual recertification tests as wildland firefighters, one from a heart attack in California and one from a CVA in Montana. One firefighter experienced a heart attack during classroom training in Virginia, and a North Carolina firefighter died of a heart attack during structural firefighting training.

Table 8. Fatalities During Training

Year	Number of Fatalities
2002	11
2001	14
2000	13
1999	3
1998	12
1997	5
1996	6

After an Incident

In 2002, 5 firefighters died after the conclusion of an emergency incident. All of the deaths were attributed to heart attacks. Three firefighters became ill as they completed paperwork at the conclusion of an EMS incident; one firefighter died in his sleep just after returning from an emergency incident; and one firefighter became ill at the scene of a structure fire, reported to the Incident Commander, went home, and was later discovered dead by friends.

Career, Volunteer, and Wildland Fatalities by Type of Duty

Figure 5 depicts career, volunteer, and wildland firefighter deaths by type of duty. Wildland career, wildland seasonal, and wildland contractor deaths were grouped together. As in past years, there were a disproportionate number of fatalities experienced by volunteer firefighters responding to and returning from alarms as compared to career and wildland firefighters. In 2002, a quarter of all volunteer firefighter deaths occurred while responding to or returning from

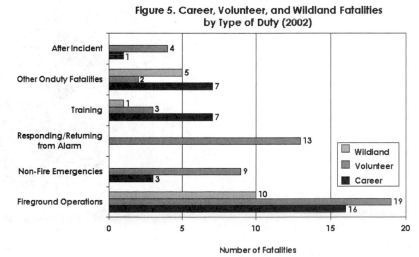

Figure 5. Career, Volunteer, and Wildland Fatalities by Type of Duty (2002)

TYPE OF DUTY

emergencies. In comparison, no career firefighter deaths and none of the wildland deaths occurred while responding or returning.

The large number of career firefighter deaths while on duty (but not involved in an incident or training activity) may be attributed to the fact that career firefighters are on duty for longer periods of time than volunteer firefighters. The onduty periods for volunteer firefighters generally are related to an emergency incident or other official functions such as training or fundraising. Some volunteer fire departments staff stations overnight (similar to a career department) but their numbers are small when compared to the total number of volunteer fire departments.

Type of Emergency Duty

In 2002, 73 firefighters died while engaged directly in the delivery of emergency services. This number includes deaths that were the result of injuries sustained on the incident scene or enroute to the incident scene, and firefighters that became ill on an incident scene and later died. It does not include firefighters who became ill or died while returning from an incident (such as a vehicle collision while returning from an incident). Figure 6 shows the number of firefighters killed in firefighting, emergency medical services, and other emergency incidents in 2002.

Fifty-eight firefighters were killed in relation to fires; 13 firefighters were killed in relation to EMS calls; and 2 firefighters were killed while engaged at a water main break in the community and assisting in the recovery after a tornado.

Figure 6. Type of Emergency Duty (2002)

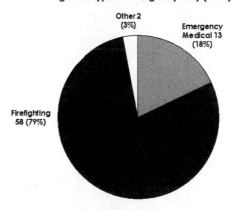

TYPE OF DUTY

CAUSE OF FATAL INJURY

The term "cause of injury" refers to the action, lack of action, or circumstances that resulted directly in the fatal injury. The term "nature of injury" refers to the medical cause of the fatal injury or illness, often referred to as the physiological cause of death. A fatal injury usually is the result of a chain of events; the first of which is recorded as the cause.

In 2002, firefighters were killed when they ran out of air or were placed into fire areas by structural collapse. The cause of the fatal injury will be listed as "collapse" and the nature of the injury will be listed as "asphyxiation" or "burns."

Similarly, if a wildland firefighter was overrun by a fire and died of burns, the cause of death will be listed as "caught/trapped" by fire progress, and the nature of death will be "burns." This follows the convention used in the NFIRS casualty reports.

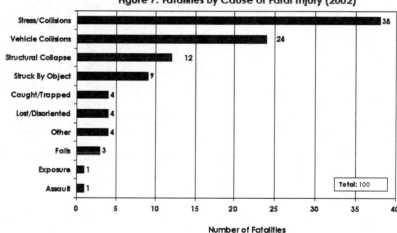

Figure 7. Fatalities by Cause of Fatal Injury (2002)

Cause	Number of Fatalities
Stress/Collisions	38
Vehicle Collisions	24
Structural Collapse	12
Struck By Object	9
Caught/Trapped	4
Lost/Disoriented	4
Other	4
Falls	3
Exposure	1
Assault	1

Total: 100

Number of Fatalities

Stress or Overexertion

The largest cause of firefighter deaths is stress or overexertion, which was listed as the primary factor in 38 of the firefighter deaths in 2002 (Table 9).

Firefighting is extremely strenuous physical work and is likely one of the most physically demanding activities that the human body performs.

Most firefighter deaths attributed to stress result from heart attacks. Of the 38 stress-related fatalities in 2002, 34 firefighters died of heart attacks and 4 died of CVA's (stroke). Ten of the 38 deaths for which the cause of the fatal injury is listed as stress/overexertion occurred during nonemergency activities.

At 38 percent this is the lowest percentage of firefighter fatalities due to stress or overexertion, and the number in any year since at least 1996.

Top of your 2003/04 Reading List: The South Beach Diet, Author: arthur Agatston, M.D. "I am convinced that preventing most heart attacks and strokes is feasible today... the majority of heart attacks and strokes can be prevented."

Table 9. Fatalities Caused by Stress or Overexertion

Year	Number	Percent of Fatalities
2002	38	38%
2001	42	41.2*
2000	45	44%
1999	54	49%
1998	42	46%
1997	40	43%
1996	47	50.0%

(* Does not include the firefighter deaths of September 11, 2001, in New York City.)

CAUSE OF FATAL INJURY

Vehicle Collisions

The second leading cause of fatal injury for firefighters who died in 2002 was vehicle collisions. This cause is usually the second most common cause of firefighter fatalities.

Five wildland firefighters were killed when the van in which they were riding was involved in a single-vehicle crash; four firefighters were killed

In 2002, a firefighter was killed in the crash of a fire pumper responding to a brush fire. The speed of the apparatus was calculated at 74 mph in a 45 mph zone just prior to the crash.

in crashes involving their personal vehicles; three firefighters were killed when their engine apparatus slipped off a road and rolled; three firefighters were killed in crashes involving tankers, one each while responding, returning from training, and returning from maintenance; one firefighter was killed when he was struck while responding on his bicycle; one firefighter was killed when his departmental vehicle was struck as he traveled to a meeting; and one firefighter was killed when the fire apparatus in which she was a passenger was involved in a crash.

Six wildland aircraft firefighters were killed in 2002 in three separate incidents. Three firefighters were killed when the wings of their aircraft separated from the fuselage and the aircraft crashed; two firefighters were killed when one wing separated from the fuselage and the aircraft crashed; and a helicopter pilot was killed while fighting a fire in Colorado.

Firefighters console one another in front of the wreckage of the tanker which was operated by Firefighter Cassandra "Sandy" Myers Billings Powell at the time of her death. The apparatus crashed onto its roof after leaving the roadway.

CAUSE OF FATAL INJURY

Structural Collapse

An unusually large number of firefighters died due to structural collapse in 2002 (Table 10).

Three multiple firefighter fatality incidents took a total of eight firefighters, three in New Jersey, three in Oregon, and two in New York. An Indiana firefighter and a Texas firefighter were killed when they were buried by structural collapses. A firefighter in North Carolina fell into a fire-involved basement after entering a structure fire, and a Pennsylvania firefighter was first pinned in a structural collapse and then killed by a subsequent massive collapse.

Twelve percent of the 2002 firefighter fatalities were related to structural collapse. This is more than twice the level of any year since 1997.

Table 10. Fatalities Caused by Structural Collapse

Year	Number	Percent of Fatalities
Year	12	12.07%
2002	4	3.9%*
2001	4	3.8%
2000	6	5.3%
1999	5	5.5%
1998	6	6.2%
1997	6	6.3%

(*Does not include WTC)

Struck by Object

Being struck by an object was the fourth leading cause of fatal firefighter injuries in 2002. There were nine deaths in this category, including four firefighters who were struck by vehicles at emergency scenes. A North Dakota firefighter was killed when he was struck by an errant fireworks shell at a community Independence Day celebration where the fire department provided the fireworks; a military firefighter in Alabama was killed when he was struck by a fire apparatus that slipped into gear and rolled forward; a New York firefighter was killed as he helped to load hose after a training exercise and was struck by a vehicle that ignored traffic control devices; a fire victim jumped and struck a firefighter at a Michigan structure fire; and a falling tree killed a wildland firefighter who was engaged in removing dangerous trees from the forest line near a road.

The photo depicts the crash scene where Captain Kim Alan Granholm of the Thomson Township/Esko Volunteer Fire Department lost his life. Only the vehicles immediately behind the fire apparatus on the shoulder were on the scene at the time of the fatal crash.

CAUSE OF FATAL INJURY

Caught or Trapped

In 2002, four firefighters were killed when they were caught or trapped. Two Florida firefighters were killed when they were overcome by fire progress during a structural firefighting training session; an Indiana firefighter was killed when he was trapped underwater during dive rescue training; and a South Dakota firefighter died when he was trapped in a molasses tank while attempting to rescue a worker.

Lost or Disoriented

Four firefighters died in 2002 when they became lost or disoriented inside a structure fire and ran out of air. Two Missouri firefighters died in a fire in a commercial building; one firefighter in Pennsylvania and one firefighter in Tennessee died when they ran out of air in residential structure fires.

Other

Four firefighters died of causes that are not categorized above. Three died of natural cardiac diseases that were not stress related, and the cause of death for one firefighter could not be determined.

Falls

Three firefighters died in 2002 as the result of falls, the same number as in 2001. An Iowa firefighter died when he fell through a ventilation hole that had been cut into the roof of a burning structure; he fell into the fire area and died of asphyxiation. Firefighters in South Dakota and Texas were killed when they fell from exterior riding positions on fire apparatus during wildland firefighting duties. One firefighter died of burns and the other firefighter was crushed by the fire apparatus as it rolled forward.

Exposure

A Maryland firefighter died of hyperthermia when he became overheated during physical fitness training during the first few days of recruit school.

Assault

A New Mexico firefighter was killed by a gunman at the scene of a structure fire. The firefighter went to assist a burn victim who was located in a house near the burning structure. As the firefighter approached the house, he was shot and killed by the burn victim whom it was learned had set the original structure fire.

From 1982 to 2002, 13 firefighters have been shot and killed by gunfire while on duty. There were two multiplefatality incidents. One of the 13 firefighters was struck when a rifle in a burning greenhouse discharged, the rest were at the hands of gunmen.
(source: Hank Przybylowicz, Line of Duty Research Service)

CAUSE OF FATAL INJURY

NATURE OF FATAL INJURY

Figure 8 shows the distribution of the 100 deaths by the medical nature of the fatal injury or illness.

Figure 8. Fatalities by Nature of Fatal Injury (2002)

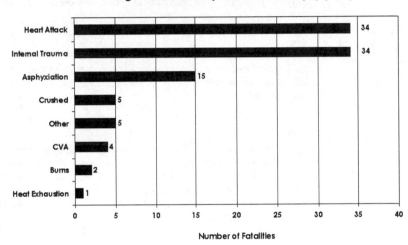

Number of Fatalities

Heart Attack

In 2002, heart attacks and internal trauma were the nature of death for 34 firefighters. Heart attacks are usually the leading nature of firefighter deaths. Figure 9 provides a detailed breakdown of heart attacks by type of duty.

Figure 9. Heart Attacks by Type of Duty (2002)

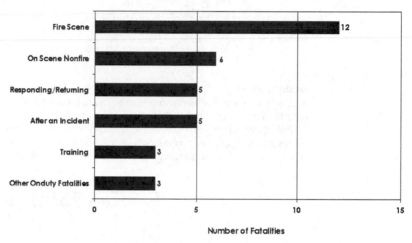

Number of Fatalities

Twelve of the heart attacks occurred at the fire scene; 6 occurred at nonfire emergencies; 5 occurred as firefighters were responding to an emergency or returning from an emergency; 5 occurred after the conclusion of an incident; and 3 occurred during other duty-related activities. Three fatal heart attacks occurred during training, down from 10 in 2001 and 7 in 2000.

Internal Trauma

In 2002, internal trauma was tied as the nature of death that took the highest number of firefighter's lives, responsible for 34 deaths (Table 11). Twenty-three of these traumatic deaths occurred where firefighters were occupants of motor vehicles. Five firefighters died in Colorado when their van was involved in a single-vehicle collision; a total of six firefighters perished in aircraft crashes; four firefighters died in crashes that involved their personal vehicles; three firefighters died when their pumper slid off a California road and rolled; three firefighters were killed in separate tanker crashes; one firefighter was killed when the pumper in which she was riding was involved in a crash; and one firefighter was killed when his fire department vehicle was struck as he drove to an association meeting.

Four firefighters were struck by vehicles at the scene of motor vehicle crashes and one firefighter was killed when he was struck at the scene of a car fire.

A New Mexico firefighter was fatally shot during a structure fire; a Texas firefighter died of traumatic injuries resulting from a structural collapse; a Colorado firefighter was killed by a falling tree; a North Dakota firefighter was killed by an exploding fireworks shell; a Pennsylvania firefighter was killed in a collision as he was responding on his bicycle; and, an Alabama firefighter was killed when a fire truck slipped into gear and struck him.

NATURE OF FATAL INJURY

Table 11. Internal Trauma Fatalities

Year	Number of Fatalities
2002	34
2001	28*
2000	36
1999	25
1998	27
1997	32
1996	32

(*Does not include WTC)

Asphyxiation

Asphyxiation was the third leading medical reason for firefighter deaths in 2002, responsible for 15 deaths (Table 12). Twelve firefighters died of smoke inhalation after being involved in structural firefights, including training. A Pennsylvania firefighter died when he was unable to breathe due to the pressure of a structural collapse; a firefighter suffocated while trying to make a rescue from a molasses tank; and a firefighter in Indiana died during dive rescue training.

Table 12. Fatalities Due to Asphyxiation

Year	Number of Fatalities
2002	15
2001	18
2000	13
1999	16
1998	15
1997	15
1996	5

NATURE OF FATAL INJURY

Crushed

In 2002, five firefighters died when they were crushed. Three New Jersey firefighters died when the residential structure that they were searching suffered a catastrophic collapse. An Indiana firefighter was crushed in the collapse of a commercial building, and a Texas firefighter was killed when he was crushed by the front wheel of a fire truck after he had fallen off the truck.

Other

Five firefighters were killed in situations where the nature of their fatal injuries do not fit into any of the categories. Three of the firefighters died of natural or hereditary heart disease; one firefighter died of respiratory failure after being exposed to smoke at a fire; and the cause of one firefighter's death could not be established.

CVA

Four firefighters died of CVA's (strokes) in 2002. One firefighter suffered a CVA while exercising at the fire station; one firefighter suffered a CVA as the result of being struck by someone who jumped from a structure fire in a residential occupancy; one firefighter suffered a CVA while performing a forestry pack test; and one firefighter suffered a CVA at the scene of a medical emergency.

Burns

Two firefighters died in 2002 where their deaths were primarily attributed to burns. Many times firefighters that die in structural fires suffer burns, but the burns are not the primary cause of death or the burns occur after the firefighter has died. A North Carolina firefighter fell into a burning basement and received severe burns. The firefighter lived for 2 days prior to dying from complications from his burns. A South Dakota firefighter was severely burned when he fell from the bed of a brush truck when the vehicle was struck. The firefighter fell into the flame front and received third-degree burns over 80 percent of his body. He lived for 5 days prior to his death from complications.

Heat Exhaustion

A Maryland firefighter was beginning his third day of recruit firefighter training. The day began with a series of physical fitness activities in extreme heat (code red) conditions. During a run returning to the training facility, the firefighter collapsed and died of hyperthermia.

NATURE OF FATAL INJURY

FIREFIGHTER AGES

Figure 10 shows the percentage distribution of firefighter deaths by age and nature of the fatal injury. Table 13 provides counts of firefighter fatalities by age and the nature of the fatal injury.

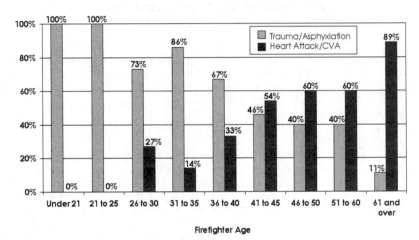

Figure 10. Fatalities by Age and Nature (2002)

Table 13. Firefighters' Ages and Nature of Fatalities

	AGE								
	Under 21	21 to 25	26 to 30	31 to 35	36 to 40	41 to 45	46 to 50	51 to 60	61+
Non-Trauma	0	0	4	1	4	7	6	12	8
Trauma	10	4	11	6	8	6	4	8	1
Total	**10**	**4**	**15**	**7**	**12**	**13**	**10**	**20**	**9**

As in most years, younger firefighters were more likely to have died as a result of traumatic injuries such as injuries from an apparatus accient or after becoming caught or trapped during firefighting operations. Stress plays an increasing role in firefighter deaths as age increases.

The median age for firefighters who suffered fatal heart attacks or CVA's on duty in 2002 was 51 years and 3 months of age. Firefighters that died from heart attacks and CVA's ranged in age from 27 to 76.

The youngest firefighter killed in 2002 was Christopher Kangas of Pennsylvania at age 14, the oldest was Fire Police Captain Harold Coons of New York at 76.

FIXED PROPERTY USE FOR STRUCTURAL FIREFIGHTING FATALITIES

There were 27 firefighter fatalities in 2002 where the firefighters became ill while on the scene or while engaged in structural firefighting and the fixed property use is known. Table 14 shows the distribution of these deaths by fixed property use. As in most years, residential occupancies accounted for the highest number of these fireground fatalities, with 21 deaths.

Table 15 shows the number of firefighter deaths in residential occupancies for the last 6 years. Residential occupancies usually account for 70 to 80 percent of all structure fires and a similar percentage of the civilian fire deaths each year. Historically, the frequency of firefighter deaths in relation to the number of fires is much higher for nonresidential structures.

Table 14. Structural Firefighting Fatalities by Fixed Property Use

Fixed Property Use	Number	Percent
Residential	21	78%
Commercial	6	22%

Table 15. Firefighter Fatalities in Residential Occupancies

Year	Number of Firefighter Fatalities
2002	21
2001	17
2000	21
1999	23
1998	17
1997	16
1996	19

FIXED PROPERTY USE

TYPE OF ACTIVITY

Figure 11 shows the types of fireground activities firefighters were engaged in at the time they sustained their fatal injuries or illnesses. This total includes all firefighting duties such as wildland firefighting and structural firefighting. In 2002 there were a total of 45 firefighter deaths on the fireground.

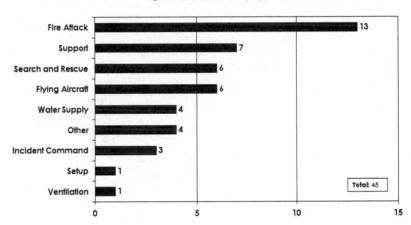

Figure 11. Fatalities by Type of Activity (2002)

Table 16. Fatalities While Engaged in Fire Attacks

Year	Number of Fatalities
2002	13
2001	13
2000	13
1999	16
1998	18
1997	21
1996	9

Fire Attack

Thirteen firefighters were killed as they engaged in direct fire attack, such as advancing or operating a hoseline at a fire scene. In years past, most fireground firefighter deaths occur while the firefighter is engaged in fire attack (see Table 16).

Three Oregon firefighters died while fighting a structural fire when a roof collapse occurred and trapped them; two New York firefighters died when a floor collapsed and they fell into a burning basement; a North Carolina firefighter also died when a floor collapsed and he fell into a burning basement; two firefighters suffered heart attacks as they fought fires; two firefighters fell from brush firefighting apparatus due to vehicle collisions and subsequently died; two firefighters died of smoke inhalation after they became lost or trapped inside burning structures; and one firefighter was pinned by a structural collapse and died while engaged in fire attack.

TYPE OF ACTIVITY

Support

Seven firefighters were killed in 2002 as they supported firefighting efforts. A Texas firefighter was killed in a structural collapse as he assisted at the scene of a structure fire; a New Mexico firefighter was killed by a gunman at the scene of a structure fire; three firefighters died as they assisted with operations at structure fires; a Minnesota firefighter died when he was struck by a vehicle at the scene of a roadside car fire; and a wildland firefighter was killed when he was struck by a falling tree.

Search and Rescue

In 2002, six firefighters died while engaged in search and rescue operations at the scene of structure fires. Three New Jersey firefighters died when the building that they were searching collapsed; two Missouri firefighters were killed as they searched a commercial building for occupants and firefighters; and a Michigan firefighter died as the result of injuries he received while attempting to rescue trapped building occupants over ground ladders - a fire victim jumped from the building and struck the firefighter.

Flying Aircraft

Six firefighters died while flying firefighting aircraft. Three firefighters died in a California crash; two died in a Colorado crash; and the pilot of a helicopter was killed when his aircraft experienced a mechanical failure and crashed.

Water Supply

Four firefighters died in 2002 while engaged in water supply duties. All four deaths were heart-related. Three occurred at structure fires and one occurred at a wildland fire.

Other

Four firefighters were killed performing activities that are not classified. Three California wildland firefighters were killed when their pumper left the roadway and rolled. They were patrolling the perimeter of a fire to protect against fire spread. A Nebraska firefighter collapsed at the scene of a brush fire and died. No further information on this incident or the circumstances surrounding the death was available.

Incident Command

Two firefighters died in 2002 as the result of heart attacks suffered as they commanded structural fires. A Massachusetts firefighter died from complications from smoke inhalation that he suffered as he was in command of operations on the scene of a structure fire.

TYPE OF ACTIVITY

Setup

An Indiana firefighter was killed as he set up equipment to begin a defensive attack on a fire in a commercial building. He was struck as the building collapsed.

Ventilation

An Iowa firefighter was killed when he fell through the roof of a fire-involved residence. The firefighter had been engaged in roof ventilation duties and fell into the fire area as he and his partner attempted to leave the roof.

TYPE OF ACTIVITY

TIME OF INJURY

The distribution of all 2002 firefighter deaths according to the time of day when the fatal injury occurred is illustrated in Figure 12 (two incident times were not reported).

Figure 12. Fatalities by Type of Injury (2002)

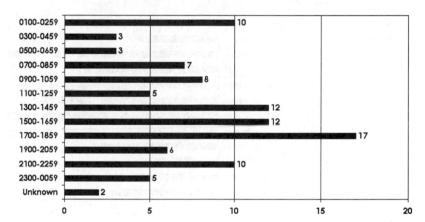

MONTH OF THE YEAR

Figure 13 illustrates firefighter fatalities by month of the year. Firefighter fatalities were highest in July due to a number of wildland firefighting deaths.

Figure 13. Fatalities by Month of the Year (2002)

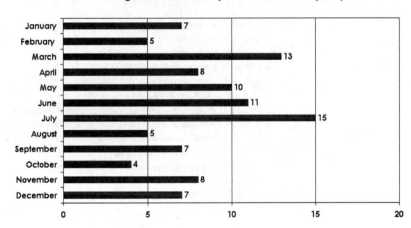

STATE AND REGION

The distribution of firefighter deaths by State is shown in Table 17. Firefighters based in 35 States died in 2002. Eight Oregon-based firefighters were killed. Five Oregon-based firefighters died in a van crash in Colorado, and three Oregon firefighters died fighting a structure fire in a commercial building.

The highest number of firefighter deaths occurred in Colorado. Nine firefighters died while on duty in Colorado in 2002. This is due to the extremely severe wildland fire season that took place in that State and two multiple firefighter fatality incidents that took a total of seven lives.

Figure 14 provides information on the ratio of firefighter fatalities per million population in each region.

Figure 14. Firefighter Fatalities by Region

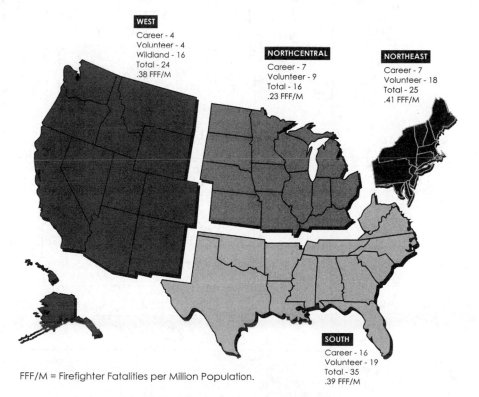

WEST
Career - 4
Volunteer - 4
Wildland - 16
Total - 24
.38 FFF/M

NORTHCENTRAL
Career - 7
Volunteer - 9
Total - 16
.23 FFF/M

NORTHEAST
Career - 7
Volunteer - 18
Total - 25
.41 FFF/M

SOUTH
Career - 16
Volunteer - 19
Total - 35
.39 FFF/M

FFF/M = Firefighter Fatalities per Million Population.

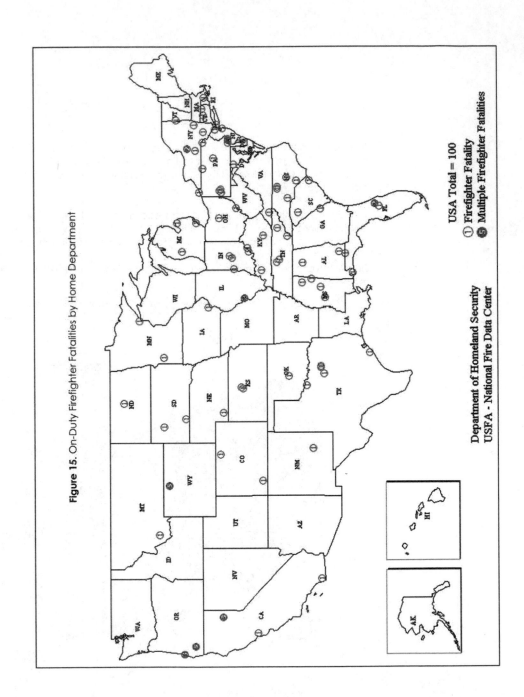

Figure 15. On-Duty Firefighter Fatalities by Home Department

USA Total = 100

① Firefighter Fatality
⑤ Multiple Firefighter Fatalities

Department of Homeland Security
USFA - National Fire Data Center

STATE AND REGION

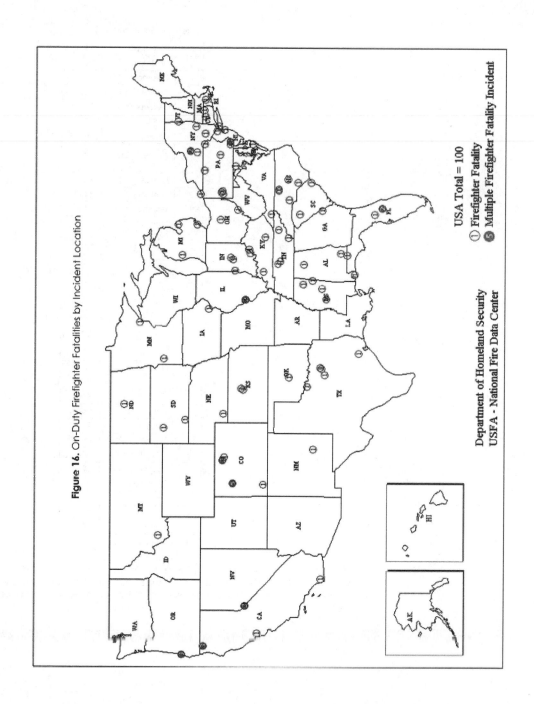

Figure 16. On-Duty Firefighter Fatalities by Incident Location

USA Total = 100
Firefighter Fatality
Multiple Firefighter Fatality Incident

Department of Homeland Security
USFA - National Fire Data Center

STATE AND REGION

Table 17. Fatalities by State

3	Alabama	3%
5	California	5%
2	Colorado	2%
3	Connecticut	3%
1	Delaware	1%
4	Florida	4%
1	Georgia	1%
1	Iowa	1%
4	Indiana	4%
2	Kansas	2%
2	Kentucky	2%
1	Massachusetts	1%
1	Maryland	1%
3	Michigan	3%
2	Minnesota	2%
2	Missouri	2%
4	Mississippi	4%
1	Montana	1%
5	North Carolina	5%
1	North Dakota	1%
1	Nebraska	1%
5	New Jersey	5%
1	New Mexico	1%
6	New York	6%
2	Ohio	2%
1	Oklahoma	1%
8	Oregon	8%
7	Pennsylvania	7%
2	South Carolina	2%
2	South Dakota	2%
5	Tennessee	5%
5	Texas	5%
1	Virginia	1%
1	Vermont	1%
5	Wyoming	5%

This list attributes the deaths according to the State in which the fire department or unit is based, as opposed to the State in which the death occurred. They are listed by those States for statistical purposes and for the National Fallen Firefighters Memorial at the National Emergency Training Center.

STATE AND REGION

ANALYSIS OF URBAN/RURAL/SUBURBAN PATTERNS IN FIREFIGHTER FATALITIES

The United States Bureau of the Census defines "urban" as a place having a population of at least 2,500 or lying within a designated urban area. Rural is defined as any community that is not urban. Suburban is not a census term but may be taken to refer to any place, urban or rural, that lies within a metropolitan area defined by the Census Bureau, but not within one of the central cities of that metropolitan area.

Fire department areas of responsibility do not always conform to the boundaries used for the census. For example, fire departments organized by counties or special fire protection districts may have both urban and rural coverage areas. In such cases, it may not be possible to characterize the entire coverage area of the fire department as rural or urban, and firefighter deaths were listed as urban or rural based on the particular community or location in which the fatality occurred.

The following patterns were found for 2002 firefighter fatalities (Table 18). These statistics are based on answers from the fire departments and, when no data from the departments were available, the data are based upon population and area served as reported by the fire departments.

Table 18. Fatalities by Coverage Area Type

Coverage Area	Fatalities
Urban/Suburban	53
Rural	33
Federal or State Parks/Wildland	14
Total	**100**

ANALYSIS OF FIREFIGHTER FATALITIES

CONCLUSION

The year 2002 was another year of unacceptable loss for the fire service in the United States. The loss of a firefighter while on duty clearly is a direct loss for the family and coworkers of the deceased firefighter. The loss, however, is also felt within the community where the loss occurred and within the international fire service family.

In early 2003, when preliminary information regarding firefighter fatalities in 2002 was known, U.S. Fire Administrator R. David Paulison said: "The American Fire Service suffered another staggering year of loss in 2002. As the Nation's first responders, firefighters across the United States put their lives in danger every day. The United States Fire Administration is committed to helping improve firefighter safety to prevent these tragedies from occurring in the future."

The fire service has an unpaid debt to each of the firefighters that died in 2002 and before -- the prevention of future firefighter deaths.

This report contains two special topics. The first looks at an issue that mainly impacts volunteer firefighters -- personal vehicle crashes. Each year, firefighters die in single vehicle crashes as they respond to the fire station or to an incident scene in their personal vehicles. The key to the prevention of these deaths is training and the development and proper administration of procedures that require a slower approach and the use of seatbelts.

The second special topic area presents some inexpensive ways to make an immediate impact on the safety of firefighters. Topics include roadside safety, medical exam information, and the prevention of deaths caused by firefighters riding on exterior positions on wildland apparatus.

SPECIAL TOPICS

PERSONALLY OWNED VEHICLES

A Tragic Record
Firefighter deaths while operating Personally Owned Vehicles (POV's) is not a recent development. In 1993, 18 firefighters died while on duty responding to and returning from alarms. Tragically, 9 of the 18 firefighters died in their POV's, representing the leading type of vehicle that firefighters are operating when they are killed.

POV-response-related fatalities have remained at alarmingly high levels. This special topic report will examine case studies from 1997 through 2002 to highlight causative factors and to offer recommendations to prevent future deaths. Only knowledge of the problem and positive steps taken in advance of the response can stem this alarming tide of firefighter fatalities.

Twenty-five firefighters died in POV crashes from 1997 through 2002.

High speed was cited as a contributing factor in almost all of the POV crashes examined for this report. One firefighter was estimated to have been going almost twice the posted 40 mile-per-hour speed limit responding back to the fire station to retrieve his Personal Protective Equipment (PPE) after having missed the initial page for a mutual-aid response to a lumber yard fire.

Lack of seatbelt use. The lack of seatbelt use still presents a big problem. At least 6 of the 25 POV-related firefighter fatalities were listed as not wearing seatbelts at the time of the accident. It is likely that more than six firefighters were not using this important piece of safety equipment but police reports on some incidents are not clear on this issue.

Rollovers. There were six POV rollover fatalities. At least four POV operators were not wearing seatbelts at the time of their rollovers. What proved to be interesting was the type of vehicle and one of the vehicle's accessories. At least three of the rollover vehicles were Sport Utility Vehicles (SUV's) and at least three of the rollover vehicles were equipped with sunroofs. In one incident, a firefighter was responding in his personally owned Chevy Tahoe to the fire station. Responding down a downgrade, the Tahoe hit a patch of ice and began to slide. The Tahoe then left the roadway and rolled over three times. During one of the rollovers, the firefighter was ejected through the sunroof and suffered a fatal head injury. Excessive speed was also a contributing factor to this crash.

Intersections. In the past 5 years, eight firefighters died in POV intersection crashes. At least three of these crashes involved the use of colored lights, whether they were courtesy blue lights or red lights and a siren on the POV's.

Alcohol use. Several POV operators had high blood alcohol concentrations and would be considered legally intoxicated in most States. Firefighters who consume alcohol should not respond to emergencies.

> *Young members of the fire department merit special attention. Younger members may use bicycles or other nonstandard means of transportation and their driving skills are not as advanced as older members.*

Head-on crashes. Head-on collisions accounted for five POV firefighter fatalities during the period. Two firefighters died when their motorcycles were involved in collisions in the performance of their duties. One junior firefighter was killed when he ran a stop sign, and was struck by a vehicle as he was responding on his bicycle.

Family members as passengers. Another troubling trend emerged during the research for this report. Three firefighters killed while driving their POV's had family members, including children, responding with them at the time of their collisions. This practice is unsafe and should not be allowed by fire department policy.

The Need for an Emergency Response? Finally, most of the firefighters killed in their POV's were responding to calls that do not fit the national definition of a "True Emergency." According to the United States Department of Transportation Emergency Vehicle Operators Course instructor's manual: "a true emergency is a situation in which there is a high probability of death or serious injury to an individual or significant property loss, and action by an emergency vehicle operator may reduce the seriousness of the situation." Most calls for assistance are **not** true emergencies, but present fire department driving attitudes do not reflect this fact -- particularly when it involves POV operation.

Recommendations

- Firefighters should be trained in safe POV driving techniques before the first response. No first responder should be allowed to respond to any call without some form of effective driver training. This training should include instruction in State motor vehicle laws related to POV use in response to emergencies, driver safety, seatbelt use, safe parking, and vehicle placement.

- POV response to emergencies should be based upon a written procedure. The procedure should include guidance on when to respond, how to respond safely, and specific response requirements regarding speed, intersections, seatbelt use, inclement weather, and limited visibility situations (dark, fog, etc.).

- POV response policies must be enforced. If an officer or a member of the fire department witnesses another member responding in a manner that is not consistent with the procedure, it must be addressed through the chain of command. If violations of the policy are not corrected, the policy is a false sense of security. Some components of the policy deserve zero tolerance with noncompliance. These could include seatbelt use, running red lights, and excessive speed.

- Vehicle-specific training should be provided. Firefighters who operate specialized or unusual POV's should receive instruction on their operating characteristics. Instruction on the operating characteristics of SUV's should be provided to those who operate them.

SPECIAL TOPICS

- **Alcohol and drug impairment.** Firefighters need to consider themselves out-of- service if they have consumed alcohol or taken drugs.

- **Control the need for emergency response.** All fire department response policies need to be structured around the Federal definition of a "True Emergency." Too many firefighters responding in their POV'S were killed while responding to nonemergency calls.

The risk to passengers must be considered. Firefighters need to consider themselves out-of-service if they have any nonfire department personnel in their POV's.

Fire Service Emergency Vehicle Safety Initiative

The *Fire Service Emergency Vehicle Safety Initiative* is a partnership effort of the USFA and the U.S. Department of Transportation (DOT)/National Highway Traffic Safety Administration (NHTSA), and the DOT/Intelligent Transportation Systems Joint Program Office.

The long-term goal of this project is to reduce the number of firefighters killed and injured responding to and returning from emergencies that account for the second highest number of onduty fatalities as well as reducing firefighter deaths and injuries from being struck while performing emergency operations on the roadway.

Further information about the National Emergency Vehicle Safety Initative may be found on the USFA Web site at www.usfa.fema.gov/inside-usfa/vehicle.cfm.

SPECIAL TOPICS

INEXPENSIVE LIFE-SAVING STEPS

There are a number of steps that can be taken by fire departments to immediately and inexpensively reduce the chances that a firefighter will be killed while on duty. The preceding special topic on POV crashes pointed out some of them. This section will illustrate some inexpensive steps that can be taken to reduce the number of firefighters who are injured and killed while on duty each year.

ROADSIDE SAFETY

As our highways and roads become more congested, the risks to firefighters operating in close proximity to traffic increase. Protection of firefighters operating on these scenes is the first step that must be addressed upon the arrival of firefighters on the scene.

Roadside Safety Basics

- Consider any moving vehicle a threat. Drivers have varying levels of capability and concentration. Until they have passed you, don't ignore them.

- Create a safety zone for firefighters to work within. Use law enforcement vehicles and fire apparatus as a shield from moving traffic.

- Wear retroreflective materials. Firefighter turnout gear contains retroreflective materials that will help firefighters to be seen while on the emergency scene. Keep PPE clean so that the retroreflective material can do its job.

- Purchase and wear safety vests. Class III vests are inexpensive and should be worn by every firefighter on the incident scene.

- Don't blind oncoming drivers. The warning lights on fire apparatus may blind or confuse approaching drivers. Turn off lights such as wig-wag headlights that cause problems. Limit the number of warning lights that are in operation.

- Give drivers early warning. Use traffic cones, law enforcement officers upstream in traffic, and flaggers to notify drivers that an emergency exists and that they need to slow down.

- Minimize the number of vehicles on the scene. When the scope of the emergency has been assessed, send units away from the scene that are not needed. Stage responding units off of the roadway in a parking lot or on-ramp until they are needed.

- Firefighters exiting an emergency vehicle need to ensure that it is safe to do so. Firefighters must watch for oncoming traffic and other hazards -- look before getting out of the vehicle.

- NFPA 1500, *Standard on Fire Department Occupational Safety and Health Program*, states that "Flourescent and retro-reflective warning devices such as traffic cones with DOT-approved retro-relective collars and DOT-approved retro-reflective signs station "EMERGENCY SCENE" (with adjustable directional arrows) and illuminated warning devices such as flares and/or other appropriate warning devices shall be used to warn

SPECIAL TOPICS

oncoming traffic of the emergency operations and the hazards to members operating at the incident." (Used with permission from NFPA, 1 Batterymarch Park, Quincy, MA 02269-9101.)

In 2002, 2 firefighters died when they arrived on the scene of roadside emergencies prior to the arrival of any emergency vehicles. These situations are extremely hazardous and require the on-scene firefighter to assume that he or she cannot be seen by drivers. The natural inclination to concentrate on the victims of the crash at the expense of personal situational awareness or a form of "tunnel vision" must be guarded against.

Free information on the safety of responders when operating at roadside emergencies is available at www.respondersafety.com.

The site includes a link to order a videotape that can be used in training and a link to a comprehensive Standard Operating Procedure (SOP) on working near moving traffic. The SOP can be customized to fit the needs of your department and can be changed to include the name of your fire department.

EXTERIOR RIDING POSITIONS

Riding on the back step or sideboards of fire apparatus has been prohibited since the very first edition of NFPA 1500. This prohibition includes the practice of firefighters riding in exposed positions on wildland or brush fire apparatus.

In 2002, 2 firefighters were killed when they were thrown from exterior riding positions on wildland or brush firefighting apparatus. In both cases, the fire apparatus was struck by another vehicle, and the firefighter was thrown off the apparatus as a result of the impact. In one case, the firefighter was thrown into the flame front and horribly burned. In the other case, the firefighter was thrown in front of the fire apparatus and was crushed as the apparatus continued to move forward.

A simple and free tactical change could prevent these types of deaths from occurring in the future.

NFPA 1500 offers some specific advice on the safe way to conduct this type of operation without the need for the firefighter to ride in an exposed position on the apparatus. The advice is contained in the annex of the standard and is therefore not a requirement of the standard, just additional information for the user of the standard.

NFPA 1500 suggests that the safe way to move along a wildland fire line is to position two firefighters to one side of the apparatus in full view (ahead) of the driver. Each firefighter should be equipped with a hoseline and the apparatus operated in noninvolved areas. As the firefighters walk and fully extinguish the fire, the apparatus follows. The driver looks out for the firefighters and the firefighters look out for the driver and the apparatus.

SPECIAL TOPICS

MEDICAL EXAMINATIONS FOR FIREFIGHTERS

NFPA 1582, *Standard on Medical Requirements for Fire Fighters* and Information for Fire Department Physicians, require annual medical evaluations for all firefighters. The evaluation includes a review of the firefighter's medical and occupational experience for the year and assessments of the firefighter's height and weight, blood pressure, and heart rate and rhythm. The annual medical evaluation can be performed by a qualified person other than the fire department physician as long as their results are reviewed by the fire department physician.

More rigorous physical examinations are required every 3 years for firefighters up to age 29, every 2 years for firefighters between the ages of 30 and 39, and every year for firefighters age 40 and over.

NFPA 1582 contains a number of resources for the fire department and the physician performing these evaluations and examinations. The standard lists physical conditions that should pose a concern and provides guidance to physicians on the physical rigors of firefighting.

A copy of the standard can be purchased for approximately $30.00, including shipping. Fire department chief or administrative officers should make sure that their fire department physician has a copy of the standard. The minimal cost involved in having a few copies of the standard available to the members of the fire department could be more than overcome with the savings realized from avoiding a single injury.

It is also a good idea to make copies of the standard available for loan to all fire department members. Loaner copies of the standard can be taken by individual members to review with their personal physicians or provided to the physicians who perform physicals associated with their full-time jobs.

Medical information is a very private subject. The confidentiality of this information is protected by a number of laws and standards. Firefighters should be able to get medical advice from someone who understands the risks and rigors of the work of a firefighter -- full-time career and volunteer firefighters included.

Firefighters console one another at the funeral for Firefighter Joshua Earley.

SPECIAL TOPICS

Appendix B

INTERNET RESOURCES OF INTEREST TO SAFETY AND HEALTH MANAGERS

The following is by no means an exhaustive list of the volumes of safety and health-related information on the Internet; however, it is by any measure, a very good start. Many, if not all, of the sites listed have links available that will lead the user to other similar sites.

American Red Cross Home Page
http://www.redcross.org

American National Standards Institute
http://www.ansi.org

American Society of Safety Engineers
http://www.asse.org

Centers for Disease Control and Prevention
http://www.cdc.gov

Chemical Emergency Preparedness and Prevention Office
http://www.epa.gov/swercepp

Commission on Fire Accreditation International
http://www.cfainet.org

Emergency Net
http://www.emergency.com

Fire Chief Magazine
http://www.firechief.com

Fire Engineering Magazine
http://www.fire-eng.com

Firefighter Close-Calls
http://www.firefighterclosecalls.com

Federal Emergency Management Agency
http://www.fema.gov

Firehouse Magazine
http://www.firehouse.com

FireNet Information Network
http://online.anu.edu.au/Forestry/fire/firenet.html

Fire Department Safety Officers Association
http://www.fdsoa.org

Fire Marshals Association of North America
http://www.nfpa.org/MemberSections/IFMA/IFMA.asp

Department of Homeland Security
http://www.dhs.gov

Industrial Safety & Hygiene News Magazine
http://www.ishn.com

Injury Control Resource Information Network
http://www.injurycontrol.com/icrin

International Association of Fire Chiefs
http://www.iafc.org

International Association of Fire Fighters
http://www.iaff.org

IAFC Volunteer Chief Officers Section
http://www.vcos.org

National Advisory Committee for Acute Exposure Guideline Levels
for Hazardous Substances
http://www.epa.gov/fedrgstr/EPA-TOX/1997/October/
Day-30/t28642.htm

National Fallen Firefighters Foundation
http://www.firehero.org

National Fire Protection Association
http://www.nfpa.org

National Association of Fire Equipment Distributors
http://www.nafed.org

National Safety Council
http://www.nsc.org

National Volunteer Fire Council
http://www.nvfc.org

National Institute for Occupational Safety and Health
http://www.cdc.gov/niosh/homepage.html

National Institute for Standards and Technology
http://www.nist.gov

National Wildfire Coordinating Group
http://www.nwcg.gov

Occupational Safety and Health Administration
http://www.osha.gov

Public Entity Risk Institute
http://www.riskinstitute.org

Responder Safety
http://www.respondersafety.com

U.S. Fire Administration
http://www.usfa.fema.gov

WiththeCommand.com
http://www.withthecommand.com

REPRINT OF THE DOCUMENT ISSUED BY THE INTERNATIONAL ASSOCIATION OF FIRE CHIEFS AND THE INTERNATIONAL ASSOCIATION OF FIRE FIGHTERS IN 1998 REGARDING THE IMPLEMENTATION OF THE TWO-IN/TWO-OUT RULE

```
┌─────────────────────────────────────────────────────┐
│          UNITED STATES DEPARTMENT OF LABOR            │
│    OCCUPATIONAL SAFETY AND HEALTH ADMINISTRATION      │
│      FIREFIGHTERS' TWO-IN/TWO-OUT REGULATION          │
└─────────────────────────────────────────────────────┘
```

The federal Occupational Safety and Health Administration (OSHA) recently issued a revised standard regarding respiratory protection. Among other changes, the regulation now requires that interior structural firefighting procedures provide for at least two firefighters inside the structure. Two firefighters inside the structure must have direct visual or voice contact between each other and direct voice or radio contact with firefighters outside the structure. This section has been dubbed the firefighters' two-in/two-out regulation. The International Association of Fire Fighters and the International Association of Fire Chiefs are providing the following questions and answers to assist you in understanding the section of the regulation related to interior structural firefighting.

1. *What is the federal OSHA Respiratory Protection Standard?*

In 1971, federal OSHA adopted a respiratory protection standard requiring employers to establish and maintain a respiratory protection program for their respiratory-wearing employees. The revised standard strengthens some requirements and eliminates duplicative requirements in other OSHA health standards.

The standard specifically addresses the use of respirators in immediately dangerous to life or health (IDLH) atmospheres, including interior structural firefighting. OSHA defines structures that are involved in fire beyond the incipient stage as IDLH atmospheres. In these atmospheres, OSHA requires that personnel use self-contained breathing apparatus (SCBA), that a minimum of two firefighters work as a team inside the structure, and that a minimum of two firefighters be on standby outside the structure to provide assistance or perform rescue.

2. *Why is this standard important to firefighters?*

This standard, with its two-in/two-out provision, may be one of the most important safety advances for firefighters in this decade. Too many firefighters have died because of insufficient accountability and poor communications. The standard addresses both and leaves no doubt that two-in/two-out requirements must be followed for firefighter safety and compliance with the law.

3. *Which firefighters are covered by the regulations?*

The federal OSHA standard applies to all private sector workers engaged in firefighting activities through industrial fire brigades, private incorporated fire companies (including the "employees" of incorporated volunteer companies and private fire departments contracting to public jurisdictions) and federal firefighters. In 23 states and 2 territories, the state, not the federal government, has responsibility for enforcing worker health and safety regulations. These "state plan" states have earned the approval of federal OSHA to implement their own enforcement programs. These states must establish and maintain occupational safety and health programs for all public employees that are as effective as the programs for private sector employees. In addition, state safety and health regulations must be at least as stringent as federal OSHA regulations. Federal OSHA has no direct enforcement authority over state and local governments in states that do not have state OSHA plans.

All professional career firefighters, whether state, county, or municipal, in any of the states or territories where an OSHA plan agreement is in effect, have the protection of all federal OSHA health and safety standards, including the new respirator standard and its requirements for firefighting operations. The following states have OSHA-approved plans and must enforce the two-in/two-out provision for all fire departments.

Alaska	Kentucky	North Carolina	Virginia
Arizona	Maryland	Oregon	Virgin Islands
California	Michigan	Puerto Rico	Washington
Connecticut	Minnesota	South Carolina	Wyoming
Hawaii	Nevada	Tennessee	
Indiana	New Mexico	Utah	
Iowa	New York	Vermont	

A number of other states have adopted, by reference, federal OSHA regulations for public employee fire fighters. These states include Florida, Illinois, and Oklahoma. In these states, the regulations carry the force of state law.

Additionally, a number of states have adopted NFPA standards, including *NFPA 1500, Standard for Fire Department Occupational Safety and Health Program*. The 1997 edition of *NFPA 1500* now includes requirements corresponding to OSHA's respiratory protection regulation. Since the NFPA is a private consensus standards organization, its recommendations are preempted by OSHA regulations that are more stringent. In other words, the OSHA regulations are the minimum requirement where they are legally applicable. There is nothing in federal regulations that "deem compliance" with any consensus standards, including NFPA standards, if the consensus standards are less stringent.

It is unfortunate that all U.S. and Canadian firefighters are not covered by the OSHA respiratory protection standard. However, we must consider the two-in/two-out requirements to be the minimum acceptable standard for safe fireground operations for all firefighters when self-contained breathing apparatus is used.

4. When are two-in/two-out procedures required for firefighters?

OSHA states that "once fire fighters begin the interior attack on an interior structural fire, the atmosphere is assumed to be IDLH and paragraph 29 CFR

1910.134(g)(4) [two-in/two-out] applies." OSHA defines interior structural fire fighting "as the physical activity of fire suppression, rescue or both inside of buildings or enclosed structures which are involved in a fire situation beyond the incipient stage." OSHA further defines an incipient stage fire in 29 CFR 1910.155(c)(26) as a "fire which is in the initial or beginning stage and which can be controlled or extinguished by portable fire extinguishers, Class II standpipe or small hose systems without the need for protective clothing or breathing apparatus." Any structural fire beyond incipient stage is considered to be an IDLH atmosphere by OSHA.

5. *What respiratory protection is required for interior structural firefighting?*

OSHA requires that all firefighters engaged in interior structural firefighting must wear SCBAs. SCBAs must be NIOSH-certified, positive pressure, with a minimum duration of 30 minutes. [29 CFR 1910.156(f)(1)(ii)] and [29 CFR 1910.134(g)(4)(iii)]

6. *Are all firefighters performing interior structural firefighting operations required to operate in a buddy system with two or more personnel?*

Yes. OSHA clearly requires that all workers engaged in interior structural firefighting operations beyond the incipient stage use SCBA and work in teams of two or more. [29 CFR 1910.134(g)(4)(i)]

7. *Are firefighters in the interior of the structure required to be in direct contact with one another?*

Yes. Firefighters operating in the interior of the structure must operate in a buddy system and maintain voice or visual contact with one another at all times. This assists in assuring accountability within the team. [29 CFR 1910.134(g)(4)(i)]

8. *Can radios or other means of electronic contact be substituted for visual or voice contact, allowing firefighters in an interior structural fire to separate from their "buddy" or "buddies"?*

No. Due to the potential of mechanical failure or reception failure of electronic communication devices, radio contact is not acceptable to replace visual or voice contact between the members of the buddy system team. Also, the individual needing rescue may not be physically able to operate an electronic device to alert other members of the interior team that assistance is needed.

Radios can and should be used for communications on the fireground, including communications between the interior firefighter team(s) and exterior firefighters. They cannot, however, be the sole tool for accounting for one's partner in the interior of a structural fire. [29 CFR 1910.134(g)(4)(i)] [29 CFR 1910.134(g)(3)(ii)]

9. *Are firefighters required to be present outside the structural fire prior to a team entering and during the team's work in the hazard area?*

Yes. OSHA requires at least one team of two or more properly equipped and trained firefighters be present outside the structure before any team(s) of firefighters enter the structural fire. This requirement is intended to ensure that the team outside the structure has the training, clothing, and equipment to protect themselves and, if necessary, safely and effectively rescue firefighters inside the structure. For high-rise operations, the team(s) would be staged below the IDLH atmosphere. [29 CFR 1910.134(g)(3)(iii)]

10. *Do these regulations mean that, at a minimum, four individuals are required, that is, two individuals working as a team in the interior of the structural fire and two individuals outside the structure for assistance or rescue?*

Yes. OSHA requires that a minimum of two individuals, operating as a team in direct voice or visual contact, conduct interior firefighting operations utilizing SCBA. In addition, a minimum of two individuals who are properly equipped and trained must be positioned outside the IDLH atmosphere, account for the interior team(s), and remain capable of rapid rescue of the interior team. The outside personnel must at all times account for and be available to assist or rescue members of the interior team. [29 CFR 1910.134(g)(4)]

11. *Does OSHA permit the two individuals outside the hazard area to be engaged in other activities, such as incident command or fire apparatus operation (for example, pump or aerial operators)?*

OSHA requires that one of the two outside person's function is to account for and, if necessary, initiate a firefighter rescue. Aside from this individual dedicated to tracking interior personnel, the other designated person(s) is permitted to take on other roles, such as incident commander in charge of the emergency incident, safety officer, or equipment operator. However, the other designated outside person(s) cannot be assigned tasks that are critical to the safety and health of any other employee working at the incident.

Any task that the outside firefighter(s) performs while in standby rescue status must not interfere with the responsibility to account for those individuals in the hazard area. Any task, evolution, duty, or function being performed by the standby individual(s) must be such that the work can be abandoned, without placing any employee at additional risk, if rescue or other assistance is needed. [29 CFR 1910.134(g)(4)(Note 1)]

12. *If a rescue operation is necessary, must the buddy system be maintained while entering the interior structural fire?*

Yes. Any entry into an interior structural fire beyond the incipient stage, regardless of the reason, must be made in teams of two or more individuals. [29 CFR 1910.134(g)(4)(i)]

13. *Do the regulations require two individuals outside for each team of individuals operating in the interior of a structural fire?*

The regulations do not require a separate two-out team for each team operating in the structure. However, if the incident escalates, if accountability cannot be properly maintained from a single exposure, or if rapid rescue becomes infeasible, additional outside crews must be added. For example, if the involved structure is large enough to require entry at different locations or levels, additional two-out teams would be required. [29 CFR 1910.134(g)(4)]

14. *If four firefighters are on the scene of an interior structural fire, is it permissible to enter the structure with a team of two?*

OSHA's respiratory protection standard is not about counting heads. Rather, it dictates functions of firefighters prior to an interior attack. The entry team must consist of at least two individuals. Of the two firefighters outside, one must perform accountability functions and be immediately available for firefighter rescue. As explained previously, the other may perform other tasks, as long as those tasks do not interfere with the accountability functions and can be abandoned to perform firefighter rescue. Depending on the operating procedures of the fire department, more than four individuals may be required. [29 CFR 1910.134(g)(4)(i)]

15. *Does OSHA recognize any exceptions to this regulation?*

OSHA regulations recognize deviations to regulations in an emergency operation in which immediate action is necessary to save a life. For fire department employers, initial attack operations must be organized to ensure that adequate personnel are at the emergency scene prior to any interior firefighting at a structural fire. If initial attack personnel find a known life-hazard situation in which immediate action could prevent the loss of life, deviation from the two-in/two-out standard may be permitted, as an exception to the fire department's organizational plan.

However, such deviations from the regulations must be exceptions and not de facto standard practices. In fact, OSHA may still issue de minimis citations for such deviations from the standard, meaning that the citation will not require monetary penalties or corrective action. The exception is for a known life rescue only, not for standard search and rescue activities. When the exception becomes the practice, OSHA citations are authorized. [29 CFR 1910.134(g)(4)(Note 2)]

16. *Does OSHA require employer notification prior to any rescue by the outside personnel?*

Yes. OSHA requires the fire department or fire department designee (i.e., incident commander) be notified prior to any rescue of firefighters operating in an IDLH atmosphere. The fire department would have to provide any additional assistance appropriate to the emergency, including the notification of on-scene personnel and incoming units. Additionally, any such actions taken in accordance with the "exception" provision should be thoroughly investigated by the fire department with a written report submitted to the fire chief. [29 CFR 1910.134(g)(3)(iv)]

17. *How do the regulations affect firefighters entering a hazardous environment that is not an interior structural fire?*

Firefighters must adhere to the two-in/two-out regulations for other emergency response operations in any IDLH, potential IDLH, or unknown atmosphere. OSHA permits one standby person only in those IDLH environments in fixed workplaces, not fire emergency situations. Such sites, in normal operating conditions, contain only hazards that are known, well characterized, and well controlled. [29 CFR 1910.120(q)(3)(vi)]

18. When is the new regulation effective?

The revised OSHA respiratory protection standard was released by the Department of Labor and published in the *Federal Register* on January 8, 1998. It is effective on April 8, 1998.

"State Plan" states have six months from the release date to implement and enforce the new regulations.

Until the April 8 effective date, earlier requirements for two in/two out are in effect. The formal interpretation and compliance memo issued by James W. Stanley, Deputy Assistant Secretary of Labor, on May 1, 1995 and the compliance memo issued by Assistant Secretary of Labor Joe Dear on July 30, 1996 establish that OSHA interprets the earlier 1971 regulation as requiring two in/two out. [29 CFR 1910.134(n)(1)]

19. How does a fire department demonstrate compliance with the regulations?

Fire departments must develop and implement standard operating procedures addressing fireground operations and the two-in/two-out procedures to demonstrate compliance. Fire department training programs must ensure that firefighters understand and implement appropriate two-in/two-out procedures. [29 CFR 1910.134(c)]

20. What can be done if the fire department does not comply?

Federal OSHA and approved state plan states must ". . . assure so far as possible every working man and woman in the Nation safe and healthful working conditions." To ensure such protection, federal OSHA and states with approved state plans are authorized to enforce safety and health standards. These agencies must investigate complaints and conduct inspections to make sure that specific standards are met and that the workplace is generally free from recognized hazards likely to cause death or serious physical harm.

Federal OSHA and state occupational safety and health agencies must investigate written complaints signed by current employees or their representatives regarding hazards that threaten serious physical harm to workers. By law, federal and state OSHA agencies do not reveal the name of the person filing the complaint, if he or she so requests. Complaints regarding imminent danger are investigated even if they are unsigned or anonymous. For all other complaints (from other than a current employee, or unsigned, or anonymous), the agency may send a letter to the employer describing the complaint and requesting a response. It is important that an OSHA (either federal or state) complaint be in writing.

When an OSHA inspector arrives, he or she displays official credentials and asks to see the employer. The inspector explains the nature of the visit, the scope of the inspection, and applicable standards. A copy of any employee complaint (edited, if requested, to conceal the employee's identity) is available to the employer. An employer representative may accompany the inspector during the inspection. An authorized representative of the employees, if any, also has the right to participate in the inspection. The inspector may review records, collect information and view work sites. The inspector may also interview employees in private for additional information. Federal law prohibits discrimination in any form by employers against workers because of anything that workers say or show the inspector during the inspection or for any other OSHA-protected safety-related activity.

Investigations of imminent danger situations have top priority. An imminent danger is a hazard that could cause death or serious physical harm immediately, or before the danger can be eliminated through normal enforcement procedures. Because of the hazardous and unpredictable nature of the fireground, a fire department's failure to comply with the two-in/two-out requirements creates an imminent danger and the agency receiving a related complaint must provide an immediate response. If inspectors find imminent danger conditions, they will ask for immediate voluntary correction of the hazard by the employer or removal of endangered employees from the area. If an employer fails to do so, federal OSHA can go to federal district court to force the employer to comply. State occupational safety and health agencies rely on state courts for similar authority.

Federal and state OSHA agencies are required by law to issue citations for violations for safety and health standards. The agencies are not permitted to issue warnings. Citations include a description of the violation, the proposed penalty (if any), and the date by which the hazard must be corrected. Citations must be posted in the workplace to inform employees about the violation and the corrective action. [29 CFR 1903.3(a)]

It is important for labor and management to know that this regulation can also be used as evidence of industry standards and feasibility in arbitration and grievance hearings on firefighter safety, as well as in other civil or criminal legal proceedings involving injury or death in which the cause can be attributed to employer failure to implement two-in/two-out procedures. Regardless of OSHA's enforcement authority, this federal regulation links fireground operations with firefighter safety.

21. *What can be done if a firefighter does not comply with fire department operating procedures for two in/two out?*

Fire departments must amend any existing policies and operational procedures to address the two-in/two-out regulations and develop clear protocols and reporting procedures for deviations from these fire department policies and procedures. Any individual violating this safety regulation should face appropriate departmental action.

22. *How can I obtain additional information regarding the OSHA respirator standard and the two-in/two-out provision?*

Affiliates of the International Association of Fire Fighters may contact:

International Association of Fire Fighters
Department of Occupational Health and Safety
1750 New York Avenue, NW
Washington, DC 20006
202-737-8484
202-737-8418 (FAX)

Members of the International Association of Fire Chiefs may contact:

International Association of Fire Chiefs
4025 Fair Ridge Drive
Fairfax, VA 22033-2868
703-273-0911
703-273-9363 (FAX)

DOES THE TWO-IN/TWO-OUT REGULATION APPLY TO VOLUNTEER FIRE DEPARTMENTS AND VOLUNTEER FIREFIGHTERS?

Yes, it is the opinion of the IAFC Volunteer Chief Officers Section (VCOS) that this regulation is applicable in all situations in which interior firefighting operations are being conducted. VCOS believes it is applicable to the vast majority of volunteer fire departments in the United States.

The VCOS supported the concept of two in/two out when it was originally proposed and VCOS still supports these new regulations since they deal directly with the safety and well-being of our firefighters.

While OSHA has stated that volunteer fire departments will not be affected by the regulations, VCOS believes that this will not be the case since 25 states align with federal OSHA. The standard also applies to private incorporated fire companies including the "employees" of incorporated volunteer companies. In addition, other non-OSHA states are giving the regulations consideration under the Environmental Protection Agency, the federal entity that typically takes the lead in non-OSHA states.

The application of the two-in/two-out rule can be argued by volunteer fire departments in many states. The reality is that two in/two out is the right thing to do for the safety of our firefighters and each volunteer fire department should seek to implement new standard operating guidelines for two in/two out.

For further information on the application of two in/two out to volunteers, contact:

Volunteer Chief Officers Section
International Association of Fire Chiefs
4025 Fair Ridge Drive
Fairfax, VA 22033-2868
703-273-0911
703-273-9363 (FAX)

FIREFIGHTER LIFE SAFETY SUMMIT
INITIAL REPORT

FIREFIGHTER LIFE SAFETY SUMMIT
INITIAL REPORT

*Courtesy of Federal Emergency
Management Agency (FEMA)*

April 14, 2004

PART I

An unprecedented gathering of the leadership of the American fire service occurred on March 10th and 11th, 2004, when more than 200 individuals assembled in Tampa to focus on the troubling question of how to prevent line-of-duty deaths. Every year approximately 100 firefighters lose their lives in the line of duty in the United States; about one every 80 hours. The first ever National Fire Fighter Life Safety Summit was convened to bring the leadership of the fire service together for two days to focus all of their attention on this one critical concern. Every identifiable segment of the fire service was represented and participated in the process.

The National Fallen Firefighters Foundation hosted the Summit as the first step in a major campaign. In cooperation with the United States Fire Administration, the Foundation has established the objectives of reducing the fatality rate by 25% within 5 years and by 50% within 10 years. The purpose of the Summit was to produce an agenda of initiatives that must be addressed to reach those milestones and to gain the commitment of the fire service leadership to support and work toward their accomplishment.

The Summit marks a significant milestone, because it is the first time that a major gathering has been organized to unite all segments of the fire service behind the common goal of reducing firefighter deaths. It provided an opportunity for all of the participants to focus on the problems, jointly identify the most important issues, agree upon a set of key initiatives, and develop the commitments and coalitions that are essential to move forward with their implementation.

Every individual who came to Tampa was already personally committed to the mission of keeping firefighters alive and all of the organizations that were represented were already on record as supporting the goal of reducing line-of-duty deaths. The Summit was designed to produce a single combined agenda for change that all of the participants, individuals and organizations could agree to support and promote. The product of their concentrated effort in Tampa will provide the foundation for a joint strategy and combined effort that will be essential to produce the desired results over the next ten years.

The Summit produced a set of initiatives that may well be regarded as radical today, however it is significant to recognize that nothing new was invented or discovered in Tampa. All of the initiatives that emerged were based on information and fundamental truths that were known long before the invitations to the Summit were issued. The gathering simply provided a forum at which those issues could be discussed openly and freely on their own merits. Some of the policies that were identified are likely to cause discomfort and controversy, however there is no arguing with the fact that the assembled leadership, who came from all segments of the fire service, concluded that these initiatives are essential to keep firefighters from dying unnecessarily.

This is the first step along a path that will require a huge commitment of energy and resources over several years. Some of the initiatives that were agreed upon will involve radical changes for the fire service. Any revolutionary movement requires committed and unwavering leadership to bring about this type of major change. The core of that leadership will come from the Summit participants who helped to shape the agenda and identify the strategies that will have to be implemented. The invited participants included key individuals who are widely recognized for their influence and leadership, some attending on their own and some as representatives of organizations that represent different sectors of the fire service. In the normal course of events these organizations often disagree on particular issues and priorities, however in Tampa the only issue on the agenda was how to keep firefighters alive and there was a very broad consensus on the efforts that are needed to accomplish that goal. As the initiatives are advanced over the next several years, the fire service will see an example of what committed leaders can accomplish when they agree to work together for an important cause.

Process

The Summit opened with an immediate emphasis on the need to take bold action to change perceptions and expectations in the fire service. The strongest words of inspiration came from NFFF Board Member Vina Drennan, who clearly reminded everyone of the pain that is felt by surviving family members and shared throughout the fire service whenever a member is killed, particularly when a life is lost in circumstances that could have been prevented. A review of historical data and statistical trends in line-of-duty deaths was presented to help the participants appreciate the range of problems and issues that must be addressed to achieve a significant reduction in fatalities.

The majority of the effort during the Summit took place in six discussion groups that focused their attention on specific domains. Groups were assembled to address:
- Structural firefighting
- Wildland firefighting
- Training and research
- Vehicle operations
- Health-wellness-fitness
- Reduction of emergency incidents and risks.

The groups were asked to produce a set of initiatives that should be undertaken to reduce line-of-duty deaths within their assigned domains. Each group was assigned two co-facilitators to lead the discussions, as well as a staff assistant to fully document the discussions. Within each of the domains the participants were asked to consider:
- Education and awareness issues

- Standards and regulations
- Specific research and technology issues
- Psychological barriers
- Leadership and personal/professional responsibility issues

The groups reported their recommendations back to the full assembly, which then produced a single consolidated set of key initiatives and implementation strategies as the product of the Summit.

The consolidated list included 16 individual initiatives:

1. Define and advocate the need for a cultural change within the fire service relating to safety, incorporating leadership, management, supervision, accountability and personal responsibility.

2. Enhance the personal and organizational accountability for health and safety throughout the fire service.

3. Focus greater attention on the integration of risk management with incident management at all levels, including strategic, tactical, and planning responsibilities.

4. Empower all firefighters to stop unsafe practices.

5. Develop and implement national standards for training, qualifications, and certification (including regular recertification) that are equally applicable to all firefighters, based on the duties they are expected to perform.

6. Develop and implement national medical and physical fitness standards that are equally applicable to all firefighters, based on the duties they are expected to perform.

7. Create a national research agenda and data collection system that relates to the initiatives.

8. Utilize available technology wherever it can produce higher levels of health and safety.

9. Thoroughly investigate all firefighter fatalities, injuries, and near misses.

10. Ensure grant programs support the implementation of safe practices and/or mandate safe practices as an eligibility requirement.

11. Develop and champion national standards for emergency response policies and procedures.

12. Develop and champion national protocols for response to violent incidents.

13. Provide firefighters and their families access to counseling and psychological support.

14. Provide public education more resources and champion it as a critical fire and life safety program.

15. Strengthen advocacy for the enforcement of codes and the installation of home fire sprinklers.

16. Make safety be a primary consideration in the design of apparatus and equipment.

Suggested Role for the National Fallen Firefighters Foundation

While developing the agenda of initiatives to reduce firefighter fatalities, the delegates were also asked to define an appropriate role for the National Fallen Firefighters Foundation in working toward the implementation of the Summit recommendations. The Foundation is committed to this mission and to taking on roles and responsibilities that it can perform more effectively than any other group. The suggestions included a list of potential approaches that could be adopted by the Foundation:

- Coalition building
- Provide advocacy for the issues
- Activism in the standards-making process
- Political and apolitical activism
- Partnerships and fundraising
- Catalyst for change
- Serve as a clearinghouse for information and data
- Develop model programs and demonstration projects
- Provide technical assistance
- Work with NIOSH on investigations
- Timely assistance to local jurisdictions
- Provide recognition for achievements and contributions
- Review performance and provide progress reports
- Public awareness and communications

The Foundation's Board of Directors will consider all of the suggested roles in the coming months and determine the ability of the Foundation to undertake each activity or project. The Board has already committed the Foundation to assume a set of responsibilities, beginning with organizing the Summit and serving as the key communicator and advocacy group for the initiatives that were developed.

In the immediate future the Foundation will join forces with the United States Fire Administration to document, publish and distribute the initiatives. The Foundation will also work with USFA to report on the progress that is achieved in implementing each of the initiatives. The Foundation has already begun to seek partners and ambassadors to move the initiatives forward and scheduled a strategy meeting to establish the immediate, mid-range and long-range priorities for a 10-year campaign. That effort should be extremely visible in the coming months.

PART II

Background discussion related to the initiatives, summarized from the Summit

Cultural Change

The most fundamental issue that was agreed upon by the Summit participants is the need for the fire service in the United States to change the culture of accepting the loss of firefighters as a normal way of doing business. This concept was reflected in several different statements that were produced by the individual discussion groups. The Summit participants unanimously declared that the time has come to change our culture and our expectations.

Within the fire service we all feel the pain with the loss of each individual firefighter, but we have come to accept the loss of more than 100 firefighters each year as a standard expectation. As long as we continue to accept this loss, we can avoid or delay making the radical and uncomfortable adjustments that will be necessary to change the outcomes. We have to convince everyone in the fire service that a line of duty death is not a standard expectation or an acceptable outcome.

Personal and Organizational Accountability

The essential cultural change has to begin with accepting personal and organizational accountability for health and safety. Every individual within the fire service has to accept a personal responsibility for health, wellness, fitness for duty, skills development, basic competencies and adherence to safe practices. The leaders and members of every fire department and every fire service organization must be accountable for the safety of their members, collectively and individually. In addition the members must be accountable to each other.

The most important and fundamental decisions relating to firefighter health and safety are made by individuals, from the top of the organizational chart to the bottom. Irresponsible behavior cannot be tolerated at any level and no external influence can overpower a failure to accept personal responsibility. The managers, supervisors and leaders within the fire service must instill and reinforce these values until they become an integral component of the culture.

Incident Management and Risk Management

This initiative incorporates a range of components that relate to our ability to safely conduct emergency operations in a high-risk environment. There is no question that fire fighters are expected to work in an environment that is inherently dangerous, however the risks and most of the specific dangers are well known. The most common causes of firefighter deaths are widely

7

recognized, along with the situations where they are most likely to occur. We have to recognize and manage the risks that apply to each situation. The essence of professionalism in the fire service is the ability to function safely and effectively within that dangerous environment. We will never be able to eliminate all of the risks, but we can be very well prepared to face most of them.

Firefighters at every level must be properly trained, equipped, organized and directed to perform their duties safely and skillfully. There must be a comprehensive structured system in place to manage incidents and risks. Company officers must be trained to supervise operations and incident commanders must be trained to manage incidents according to standard principles and practices.

Firefighters must be prepared to function competently in a wide range of situations, including critical events that can involve unanticipated dangers and immediate risks to their own survival. Several areas were identified for special emphasis, including mayday and rapid intervention procedures, air management and preventing disorientation in zero-visibility conditions.

Risk management involves identifying the situations where predictable risks are likely to be encountered and making decisions that will reduce, eliminate or avoid them.

Realistic risk management applies at every level within the fire service, from the decisions made by individual firefighters and company officers to the actions of incident commanders and senior officers who have specific responsibilities for evaluating and managing risks. We fail to act professionally when we recognize a risk and choose to do nothing about it.

The willingness of firefighters to risk their own lives to save others must never be used as an excuse to take unnecessary risks. Firefighters are highly respected for being willing to risk their own lives to save others, but that cannot justify taking unnecessary risks in situations where there is no one to save and nothing to be gained. In too many cases firefighters lose their lives while trying to save property that is already lost or to rescue victims who are already dead. While these efforts are valiant, they are also futile. Individual firefighters who take unnecessary risks, or fail to follow standard safety practices, endanger their own lives as well as the lives of other fire fighters who are depending on them or who might have to try to rescue them.

Right to Stop Unsafe Procedures

The Summit participants identified the fundamental right and responsibility of firefighters to stop unsafe procedures as a key issue. To many members of the

NFFF/USFA   **Firefighter Life Safety Summit Initial Report**

fire service, who have been indoctrinated with a traditional sense of unquestioning discipline, this could be an uncomfortable concept. The underlying principle is that an individual who recognizes an unsafe situation must take action to prevent an accident from occurring. Under this operational concept, any firefighter who believes that a situation is unsafe, or could be unsafe, has both the right and the responsibility to stop the action while an evaluation is made.

The justification for this policy in non-emergency activities, particularly training situations, is easily understood. The application of the same concept to emergency operations could be more difficult for some individuals to accept. In too many cases the investigation of a fatal accident determines that an unsafe situation was recognized, but no action was taken to change or reconsider the plan. There are very few situations, even during emergency operations, where a brief hesitation to re-evaluate a potentially dangerous plan of action would have serious negative consequences. On the contrary, experience has shown that many lives could have been saved by taking a few extra seconds to stop and think.

This policy does not mean that no action can be taken at an emergency scene that exceeds the comfort level of any individual. The obligation attached to the policy is to pause long enough to determine if it is reasonably safe to continue. In many cases an officer will have to make a very rapid assessment of the situation and decide whether to continue or change the plan.

During emergency incidents there is one key question that we need to ask ourselves at regular intervals: "Are the results we are trying to accomplish worth the risks we are taking with our people?" The answer to this question should dictate actions by commanders and firefighters.

Mandatory National Standards for Training and Qualifications

The Summit participants reached the conclusion that the time has come to apply a mandatory national uniform system of training and qualification standards for all firefighters. This system would establish mandatory training, education and performance requirements, based on the duties an individual is expected to perform, regardless of their status within the fire service or the type of organization. The roles and responsibilities of firefighters at different levels and in different operating environments would be clearly defined.

The basic system of professional qualifications standards already exists, however their current applicability depends on state and local jurisdictions, individual organizations, seniority, whether a firefighter is a career, part-time or volunteer member of the fire service, and many other factors. The summit delegates agreed that we must move toward a system where the same standards would be applied to anyone who performs a given role within the fire service. The applicable standards must be appropriate and realistic for the functions the individual is expected to perform, however there should not be standards that apply to some individuals and not to others.

In reference to the existing professional qualification and certification systems, the need for periodic recertification was identified as a priority. Today, in many cases, an individual can be certified at a particular level and retain that certification for life, with no requirements for continuing education, refresher classes, performance testing or skills evaluation. The established systems for emergency medical practitioner certification were suggested as a model.

Mandatory Medical and Physical Fitness Standards

Mandatory requirements for medical examinations and physical fitness standards should also be implemented for all firefighters, based on the duties and functions they are expected to perform. Medical surveillance should be increased before strenuous physical activity and during activities where firefighters are expected to operate at extreme levels of exertion and endurance.

An increased emphasis on health, wellness and fitness is essential to reduce the number of deaths resulting from heart attacks and other cardiovascular causes. Statistics suggest that the most significant reductions in line of duty deaths are likely to be achieved through increased medical surveillance and physical fitness programs. The need for improvements in this area is most pronounced in the volunteer fire service, where the rate of fatalities due to heart attacks and other cardiovascular causes is now much higher than within the career service. This is a reversal of the situation that existed twenty years ago, when there were more cardiovascular deaths among career firefighters than volunteers.

National Research Agenda and Data Collection

The discussion of on-going research efforts as well as priorities for additional research projects pointed to the need for a national research agenda and budget for the fire service. Several different areas were listed as operational research priorities, from building construction to communications systems, leadership, management practices, and decision-making. A separate research agenda relating to the full spectrum of health, wellness and physical fitness issues was also identified, including psychological and physiological stress, cardiovascular function, oncology and biofeedback. The need for more comprehensive data collection and sharing to identify problem areas and support research was also recognized.

The researchers who were present described their current projects, as well as asking for guidance on the most pressing areas to explore. Several research projects are being conducted by different organizations on a variety of subjects relating to health and safety, however most of their funding and support come from sources outside the fire service and the projects are often influenced by other priorities and agendas. The fire service needs a process to identify its own research priorities and coordinate efforts, as well as a dedicated source of funding that can be allocated to the most important research projects. The delegates also recognized the challenge of applying many of the current research findings to real world situations.

The fire service is benefiting from crossover adaptations of technologies that were developed for other purposes; however only limited funding is available to develop new technology specifically for fire service applications. The delegates emphasized the need to make the most use of technological advancements that are available to reduce the risks of emergency operations and training exercises. Cost is often mentioned as a barrier that keeps fire departments from adopting technological improvements that could improve the safety of operations. If a technological solution is available to eliminate a known risk, the cost must be considered in relation to the consequences of not making the investment to protect the lives of firefighters.

Fatality Investigations

The need for more consistent and comprehensive investigations and data collection to analyze the causes of fire fighter fatalities was also identified as an important priority. The delegates noted that the existing NIOSH fatality investigation program needs to be expanded and that every fatality should be thoroughly investigated and documented by a team of investigators who are qualified to examine all of the pertinent factors. The same type of investigation should be conducted for serious injuries and near-miss incidents to focus on preventing future occurrences. The need for autopsy results based on a standard protocol for every line of duty death was reinforced.

Safety in Apparatus and Equipment Design

Improvements in the design and construction of apparatus and equipment are needed to address a long list of concerns. The areas discussed ranged from breathing apparatus improvements to reducing the risk of rollover accidents involving tanker apparatus. Several on-going research and development projects were described and the manufacturers and suppliers who were present noted numerous suggestions for potential improvements.

Several safety issues relating to the danger of firefighters being struck by vehicles while operating on roadways were discussed. A comprehensive approach involving apparatus positioning, emergency lighting, warning signs and traffic control devices, high visibility protective clothing, coordination with police agencies and public education was identified as a high priority for nationwide training and implementation.

Incentives and Grants Tied to Safe Practices

Several discussion groups identified the need to create a direct link between reinforcing appropriate safety policies and practices and the availability of state and federal grant funds for fire departments. There was unanimous support for the basic concept of using grant funds to leverage improvements in heath and safety programs. The suggestions ranged from making special incentive grants to fire departments in order to implement safe practices to making compliance with health and safety standards a condition for receiving funds under existing grant programs.

The Summit participants also supported the idea of officially recognizing achievements and significant improvements in heath and safety. Several individual projects and organizations were suggested as already deserving recognition.

Response Policies

Another need for cultural change was identified in relation to emergency vehicle operations and response policies. The group that examined this area noted that an average of 10 firefighters are killed each year in vehicle accidents while responding to emergency incidents and an even greater number of civilians die in collisions involving responding emergency vehicles.

Many of the emergency response deaths result from excessive speed and unsafe driving, which can be related to the perception that the urgency of the mission justifies an elevated level of risk to the emergency responders and everyone else on the streets. In too many cases the risks that are created en route are greater than the dangers of the situation itself. The cultural change must be based on recognizing that firefighters cannot save lives or property at the scene of an emergency incident unless they arrive safely and there is no justification for causing more harm en route than they can prevent when they arrive.

The need for standard policies governing emergency response was identified, possibly through the development of a new NFPA standard or by adding to the scope of an existing standard. These policies should determine when emergency response is and is not appropriate and include specific policies relating to responding in privately owned vehicles.

This cultural change has to begin with the enforcement of existing safe driving protocols by leaders and supervisors, as well as the mandatory use of seatbelts by all firefighters. The delegates noted that in many cases firefighters do not use seatbelts that are provided in their vehicles, in spite of NFPA standards, departmental regulations and state laws. The failure to enforce and to follow these existing and basic safety procedures was highlighted as evidence of the urgent need for cultural change.

The delegates recommended the adoption of a special classification of driver's license for emergency vehicle operators, similar to the existing commercial driver's license program. Instead of providing special exemptions for emergency vehicle operators, regulatory authorities should establish strict training and testing requirements, including medical clearances and periodic review of driving records. The highest standards should be applied to emergency vehicle operators.

Protocol for Violent Incidents

The Summit delegates identified the need for special protocols and policies in several areas, including a protocol for responding to violent incidents. This recognized the increasing exposure of firefighters to violent crimes and situations, up to and including terrorist attacks.

Counseling and Psychological Support

Based largely on experiences related to 9-11 and FDNY, counseling and psychological support programs were identified as high priorities for increased attention. Some of the accepted concepts and programs have been found to be inadequate or counterproductive. A combination of research efforts and large-scale practical experience will be required to make the necessary changes and improvements to these programs.

Public Education

The potential impact of public education on firefighter safety was addressed in a variety of contexts. Public fire and life safety education was identified as a strategy to reduce fires and the resulting level of risk to firefighters. In a similar manner, programs that are designed to prevent injuries and teach citizen CPR will reduce the risks encountered when responding to emergency medical and rescue incidents. Teaching drivers how to react appropriately when they encounter an emergency vehicle or when approaching an incident on a roadway will reduce another component of risk to firefighters.

Fire Prevention Codes and Residential Sprinklers

One of the most productive strategies for reducing the risk of fire fighter fatalities is to reduce the frequency of fires and emergency incidents. A comprehensive effort to increase fire service activism in fire prevention, code development and code enforcement should have a direct impact on reducing the exposure of fire fighters to dangerous situations. Efforts to promote the installation of automatic sprinkler systems in all new residential construction will have a profound impact on future fire rates. Addressing the problem of juvenile fire setters will have a positive impact on firefighter safety as well as general public safety. The delegates endorsed placing an increased emphasis on prevention as a long-term strategy to reduce fire fighter fatalities.

FIREFIGHTER LIFE SAFETY INITIATIVES

1. Define and advocate the need for a cultural change within the fire service relating to safety incorporating leadership, management, supervision, accountability, and personal responsibility.

2. Enhance the personal and organizational accountability for health and safety throughout the fire service.

3. Focus greater attention on the integration of risk management with incident management at all levels, including strategic, tactical, and planning responsibilities.

4. All firefighters must be empowered to stop unsafe practices.

5. Develop and implement national standards for training, qualifications, and certification (including regular recertification) that are equally applicable to all firefighters based on the duties they are expected to perform.

6. Develop and implement national medical and physical fitness standards that are equally applicable to all firefighters, based on the duties they are expected to perform.

7. Create a national research agenda and data collection system that relates to the initiatives.

8. Utilize available technology wherever it can produce higher levels of health and safety.

9. Thoroughly investigate all firefighter fatalities, injuries, and near misses.

10. Grant programs should support the implementation of safe practices and/or mandate safe practices as an eligibility requirement.

11. National standards for emergency response policies and procedures should be developed and championed.

12. National protocols for response to violent incidents should be developed and championed.

13. Firefighters and their families must have access to counseling and psychological support.

14. Public education must receive more resources and be championed as a critical fire and life safety program.

15. Advocacy must be strengthened for the enforcement of codes and the installation of home fire sprinklers.

16. Safety must be a primary consideration in the design of apparatus and equipment.

ACRONYMS

ADA	Americans with Disabilities Act	HBV	Hepatitis B virus
ADEA	Age Discrimination in Employment Act	HIPAA	Health Insurance Portability and Accountability Act
AFGP	Assistance to Firefighters Grant Program	HIV	Human immunodeficiency virus
ANSI	American National Standard Institute	HMEP	Hazardous Materials Emergency Preparedness
ASTM	American Society of Testing and Materials	HSO	Health and safety officer
BLEVE	Boiling liquid expanding vapor explosion	HSSP	Homeland Security Standards Panel
BSI	Body substance isolation	IAFC	International Association of Fire Chiefs
CBRN	Chemical, biological, radiological, nuclear	IAFF	International Association of Fire Fighters
CBRNE	Chemical, biological, radiological, nuclear, or explosive	IC	Incident commander
CDC	Centers for Disease Control and Prevention	IDLH	Immediately dangerous to life and health
CDL	Commercial driver's license	IMS	Incident management system
CFR	Code of Federal Regulations	ISO	Incident safety officer
CISM	Critical incident stress management	IV	Intravenous
CVD	Cardiovascular disease	LEPC	Local emergency planning committee
EMS	Emergency medical services	MVA	Motor vehicle accident
EPA	Environmental Protection Agency	NAEVT	National Association of Emergency Vehicle Technicians
EVOC	Emergency vehicle operator courses	NBC	Nuclear, biological, or chemical
FACE	Fatality Assessment and Control Evaluation	NFIRS	National Fire Incident Reporting System
FIRE	Fire Investment and Response Enhancement	NFPA	National Fire Protection Association
GIS	Geographic Information System	NIOSH	National Institute for Occupational Safety and Health
GPS	Global positioning system	NIST	National Institute for Standards and Technology
HAZWOPER	Hazardous Waste Operations and Emergency Response	NVFC	National Volunteer Fire Council

OSHA	Occupational Safety and Health Administration
OVC	Organic vapor canister
PAR	Personal accountability reports
PASS	Personal alert safety system
PIA	Postincident analysis
PM	Preventive maintenance
PPE	Personal protective equipment
PSOB	Public Safety Officer's Benefit Program
RAID	Recognize, approach, identify, decide
RCFP	Rural Community Fire Protection
RIC	Rapid intervention company

RIT	Rapid intervention team
RPD	Recognition-primed decision
SCBA	Self-contained breathing apparatus
SLFD	St. Louis Fire Department
SOG	Standard operating guidelines
SOP	Standard operating procedure
THL	Total heat loss
TPP	Thermal protective performance
USFA	United States Fire Administration
VCOS	Volunteer Chief Officer's Section
VFA	Volunteer Fire Assistance
WMD	Weapons of mass destruction

GLOSSARY

Aerobic fitness A measurement of the body's ability to perform and utilize oxygen.

Alternative duty programs Sometimes called light duty or modified duty, these programs allow an injured employee to return to work, with restrictions, for some period of time while recuperating.

Back drafts Occur as a result of burning in an oxygen-starved atmosphere. When air is introduced, the superheated gases ignite with enough force to be considered an explosion.

Boiling liquid expanding vapor explosions (BLEVES) Occur when heat is applied to a liquified gas container and the gas expands at a rapid rate while the container is weakened by the heat. When the container fails, a BLEVE is said to occur.

Bloodborne pathogen Disease carried in blood or blood products.

Body composition A measure to show the percentage of fat in the body; there are certain published parameters for what is considered average or normal.

Body substance isolation (BSI) An infection control strategy that considers all body substances potentially infectious and isolates all body substances that might be infectious.

Cardiovascular fitness Fitness levels associated with the cardiovascular system, including the heart and circulatory system.

Civil disturbances Uprisings of civilians that often lead to hostile acts against law enforcement and emergency responders.

Code of Federal Regulations (CFR) The document that contains all of the federally promulgated regulations for all federal agencies.

Cognitive Skills learned through a mental learning process as opposed to practical learning.

Consensus standards Standards developed by consensus of industry or subject area experts, which are then published and may or may not be adopted locally. Even if not adopted as law, these standards can often be used as evidence for standard of care.

Crew unity The concept that a fire company or unit shall remain together in a cohesive, identifiable working group to ensure personnel accountability and the safety of all members. A company officer or unit leader shall be responsible for the adequate supervision, control, communication, and safety of members of the company or unit.

Critical incident stress Stress associated with critical incidents, such as the injury or death of a coworker or a child.

Critical incident stress management (CISM) A process for managing the short- and long-term effects of critical incident stress reactions.

Demobilization The process of returning personnel, equipment, and apparatus after an emergency has been terminated.

Fire brigades The use of trained personnel within a business or at an industry site for firefighting and emergency response.

First in/last out The common approach often used at an emergency scene. Basically, the first arriving crews are generally the last to leave the scene.

Flashover A sudden, full involvement in flame of materials and gases within a room.

Flexibility The range of motion in a joint, which depends on the extensibility of soft tissues.

Freelancing Occurs when responders do not follow the incident plan at a scene and do what they want on their own. A failure to stay with assigned group.

Frequency How often a risk occurs or is expected to occur.

Goals Broad statements of what needs to be accomplished.

Haddon matrix A 4 × 3 matrix used to help analyze injuries and accidents in an attempt to determine processes designed to reduce them.

Heat transfer The transfer of heat through conduction, convection, radiation, and direct flame contact.

Hits Number of documents found when searching for a key word or phrase using a search engine.

Immediately dangerous to life and health (IDLH) Used by several OSHA regulations to describe a process or an event that could produce loss of life or serious injury if a responder is exposed or operates in the environment.

Incident management system (IMS) An expandable management system for dealing with a myriad of incidents to provide the highest level of accountability and effectiveness. Limits span of control and provides a framework of breaking the big job down into manageable tasks.

Interior structural firefighting The physical activity of fire suppression, rescue, or both, inside buildings or enclosed structures that are involved in a fire beyond the incipient stage.

International Association of Fire Chiefs (IAFC) Organization of fire chiefs from the United States and Canada.

International Association of Fire Fighters (IAFF) Labor organization that represents the majority of organized firefighters in the United States and Canada.

Links Used on Internet pages so that each page may be directly tied to a page on another Internet site.

Matrix A chart used to categorize actions or events for analysis.

Muscular endurance The ability of the muscle to perform repeated contraction for a prolonged period of time; the ability of the muscle to persist.

Muscular strength The maximal amount of force a muscle or group of muscles can exert in a single contraction; the ability to apply force.

Mushrooming A situation in which heat and gases accumulate at the ceiling or top floor of a multistory building then back down. Can be prevented with vertical ventilation.

National Association of Emergency Vehicle Technicians (NAEVT) An organization that bestows professional certification in many areas for persons involved in emergency service vehicle maintenance.

National Fire Incident Reporting System (NFIRS) The uniform fire incident reporting system for the United States; the data from this report is analyzed by the United States Fire Administration.

NFPA 1403 The National Fire Protection Association's consensus standard, *Live Fire Training Evolutions*.

NFPA 1500 The National Fire Protection Association's consensus standard *Fire Department Occupational Safety and Health Program*.

Occupational disease An abnormal condition, other than an occupational injury, caused by an exposure to environmental factors associated with employment.

Occupational injury An injury that results from exposure to a single incident in the work environment.

Occupational Safety and Health Administration (OSHA) Federal agency tasked with the responsibility for the occupational safety of employees.

OSHA 300 Log of Work-Related Injuries and Illnesses.

OSHA 300-A Summary of Work-Related Injuries and Illnesses.

OSHA 301 The Injury and Illness Incident Report.

Outcome evaluation An evaluation that answers the question, "Did the program meet the expected goals?"

Peer defusing The concept of using a trained person from the same discipline to talk to an emergency responder after a critical incident occurs as a means to allow the responder to talk about his or her feelings about the event in a nonthreatening environment.

Personal alert safety system (PASS) A device that produces a high-pitched audible alarm when the wearer becomes motionless for some period of time; useful for attracting rescuers to a downed firefighter.

Personnel accountability reports (PARs) A verbal or visual report to incident command or to the accountability officer regarding the status of operating crews. Should occur at specific time intervals or after certain tasks have been completed.

Postincident analysis (PIA) A critical review of the incident after it occurs. The postincident analysis should focus on improving operational effectiveness and safety.

Preventive maintenance (PM) program An ongoing program for maintenance on vehicles. Designed to provide routine care, oil changes and the like, as well as catch minor problems before they become major ones.

Process evaluation The evaluation of the various processes associated with a program or task that is ongoing.

Rapid interview companies (RIC) An assignment of a group of rescuers with the sole purpose of rapid deployment to reports of operating personnel in trouble or missing.

Regulations Rules or laws promulgated at the federal, state, or local level with a requirement to comply.

Rehabilitation The group of activities that ensures responders' health and safety at an incident scene. May include rest, medical surveillance, hydration, and nourishment.

Risk control A common approach to risk management in which measures and processes are implemented to help control the number and the severity of losses or consequences of risk to the organization.

Risk management A process of using the available resources of an organization to plan and direct the activities in the organization so that detrimental effects can be minimized.

Risk transfer The process of letting someone else assume the risk; for example, buying auto insurance transfers the consequences of an accident to the insurance company.

Risks The resultant outcome of exposure to a hazard.

Rollover The rolling of flame under the ceiling as a fire progresses to the flashover stage.

Search engines Programs on the Internet that allow a user to search the entire Internet for key words or phrases.

Self-contained breathing apparatus (SCBA) An atmosphere-supplying respirator for which the breathing air source is designed to be carried by the user.

Severity How severe the result is when a risk occurs.

Standard of care The concept of what a reasonable person with similar training and equipment would do in a similar situation.

Standard operating procedures (SOPs) Sometimes called standard operating guidelines, these are department-specific operational procedures, policies, and rules made to assist with standardized actions at various situations.

Standardized apparatus Apparatus that has exactly the same operation and layout of other similar apparatus in a department. For example, all of the department's pumpers would be laid out the same, operate the same, and have the same equipment. Useful for the situations when crews must use another crew's apparatus.

Standards Often developed through the consensus process, standards are not mandatory unless adopted by a governmental authority.

Stress The body's reaction to an event. Not all stress is bad; in fact, some level of stress is necessary to get a person to perform, for example, the stress associated with a report that is due is often the motivating factor in doing it.

Thermal Protective Performance rating The amount of protection against both convective and radiant heat that your composite gear outer shell, thermal liner, and moisture barrier should be able to provide in the event of a flashover.

Tort Liability An action that harms another person or group and can occur when a person or group of people act, or fail to act, and thus harm another directly or indirectly.

Total Heat Loss (THL) The measurement of an ensemble's heat stress reduction capability. The higher the THL value, the greater the benefit for the firefighter.

Unified command Used in the IMS when two or more jurisdictions or agencies share incident command responsibilities but do so in conjunction with each other.

United States Fire Administration Agency under the Federal Emergency Management Agency that directs and produces fire programs, research, and education.

Universal precautions Term used to describe the practice of treating all patients as if they carried an infectious disease, in terms of body fluid contamination.

INDEX

W

Wellness and fitness, 14, 83–86
  emotional fitness, 86
  medical fitness, 83–84
  physical fitness, 84–86

Wildfires, 100–101
  safety tips for, 101
Wind chill index, *114*
Workers' compensation insurance, 58, 208
Workforce 2000, 224